Ice Sheets and Climate

Ice Sheets
and Climate

by

J. Oerlemans

and

C. J. van der Veen

Institute of Meteorology and Oceanography,
State University of Utrecht, The Netherlands

D. REIDEL PUBLISHING COMPANY

A MEMBER OF THE KLUWER ACADEMIC PUBLISHERS GROUP

DORDRECHT / BOSTON / LANCASTER

Library of Congress Cataloging in Publication Data

Oerlemans, J. (Johannes), 1950–
    Ice sheets and climate

        Bibliography: p.
        Includes index
        1. Ice sheets. 2. Climatology-Mathematical models.
I. Veen, C.J. van der (Cornelis J.), 1956   . II. Title
QC981.8T23037   1984   551.5   83-27020
ISBN 90-277-1709-5

Published by D. Reidel Publishing Company,
P.O. Box 17, 3300 AA Dordrecht, Holland.

Sold and distributed in the U.S.A. and Canada
by Kluwer Academic Publishers,
190 Old Derby Street, Hingham, MA 02043, U.S.A.

In all other countries, sold and distributed
by Kluwer Academic Publishers Group,
P.O. Box 322, 3300 AH Dordrecht, Holland.

CONTENTS

PREFACE

Climate modelling is a field in rapid development, and the study of cryospheric processes has become an important part of it.

On smaller time scales, the effect of snow cover and sea ice on the atmospheric circulation is of concern for long-range weather forecasting. Thinking in decades or centuries, the effect of a CO_2 climatic warming on the present-day ice sheets, and the resulting changes in global sea level, has drawn a lot of attention. In particular, the dynamics of marine ice sheets (ice sheets on a bed that would be below sea level after removal of ice and full isostatic rebound) is a subject of continuous research. This interest stems from the fact that the West Antarctic Ice Sheet is a marine ice sheet which, according to some workers, may be close to a complete collapse.

The Pleistocene ice ages, or glacial cycles, are best characterized by total ice volume on earth, indicating that on large time scales (10^4 to 10^5 yr) ice sheets are a dominant component of the climate system. The enormous amount of paleoclimatic information obtained from deep-sea sediments in the last few decades has led to a complete revival of interest in the physical aspects of the Pleistocene climatic evolution.

So ice sheets form an active part of the climate system, and it has recently been recognized that climate modelling should involve some modelling of ice sheets, at least when one is interested in longer time scales. However, here a gap exists between the fields of glaciology and climate research. Models used in glaciology are generally designed for detailed local studies rather than for global climate studies.

In this book we try to bridge this gap. It has been written from the point of view of a climate modeller, wishing to include ice sheets in his/her model studies. One may therefore consider this text as a chapter out of a book on climate dynamics.

We tried to write this book in the 'spirit of quantification'. Although in this interdisciplinary field it is extremely difficult to construct models that are more or less 'complete', simple models picking out a particular mechanism can be very useful, and certainly better than mere speculation (so frequently encountered in climate research).

We realize that this book relies heavily on our own work, and that it probably does not give credit to all scientists working

ix

in this field. However, our primary goal was to write a short
text dealing with the specific role ice sheets play in the
climate system, at a level suitable for post-graduate courses.
For a comprehensive discussion on how ideas about ice sheets
evolved, and on what is known from field work on present-day
and former ice sheets, the reader is referred to Denton and
Hughes (1981). This work also contains an extensive list of
references.

The text has been organized in such a way that it can be studied
in two ways. One is to go straight through, of course, but it is
also possible to skip chapters 3 to 7. Chapters 1 and 2 deal
with the global climate system and how ice sheets affect it, in
general terms. In chapters 8 to 12 the discussion deepens, and
here many results are used that were obtained from simple and
more complicated numerical models. The development of these
'tools' is in fact discussed in chapters 3 to 7.

We hope that this book will prove to be helpful in courses on
climate dynamics, and that it may stimulate glaciologists to see
ice sheets as an aspect of climate, and climatologists to
consider ice sheets as an active component of the climate system.

ACKNOWLEDGEMENTS

When looking back at the process of writing this text, it appears difficult to see whose ideas, polite criticism, and/or unpleasant remarks were most valuable. We certainly learned a lot from discussions at international meetings, remarks from students, and criticism from reviewers that handled the papers on which parts of this book are based. We really appreciate their interest.

We are particularly indebted to Cor Schuurmans and Hendrik van Aken for carefully reading the manuscript. No text is free of errors, but without their help there would have been much more misprints and unclear formulations. We thank them for their efforts.

We further acknowledge permission from:
the International Glaciological Society to reproduce from the
  Journal of Glaciology Figure 11.5,
Friedr. Vieweg & Sohn to reproduce Figures 2.7 and 2.8,
the American Meteorological Society to reproduce Figures 2.7
  and 2.8,
Academic Press to reproduce Figure 1.9,
Prentice Hall to reproduce Figure 2.4,
Munksgaard Intern. Publ. to reproduce Figures 5.5, 5.6 and 7.3,
Prof. J. Imbrie to reproduce Figure 10.3,
the Publishers of Nature to reproduce Figures 11.4, 11.7, 11.8,
  11.9 and 11.10,
the Intern. Ass. Hydrological Sci. to reproduce Figure 11.12.

C.J. van der Veen is supported by the Netherlands Organization for the Advancement of Pure Research (Z.W.O.).

1. THE GLOBAL CLIMATE SYSTEM

The intention of this book is to discuss the role played by
ice sheets in the climate system. This implies that the main
interest is in the larger scales, both in time and space.
We therefore start with a sketch of the global climate system.

1.1 The global energy balance

Differences in solar heating are the driving force of the climate
system. On the avarage, the radiation budget is positive at
lower latitudes and negative at higher latitudes, and a poleward
flux of energy is thus required to create a state of balance.
This flux is accomplished by large-scale atmospheric motions and
ocean currents. Considering the global climate system as a
whole, however, dynamical energy fluxes do not appear in the
energy balance. We will first discuss the global mean system
in a long-term mean state.
    The sun radiates at an effective temperature of about 5700 K,
whereas a typical temperature of the climate system is 270 K.
As a consequence, solar and terrestrial radiation are in
different spectral intervals (solar: 0.2 to 3 µm; terrestrial:
3 to 100 µm) and it should thus be expected that the
absorptivity of the atmosphere will be different for solar
and terrestrial radiation.
    A natural boundary in the atmosphere, separating regions
of different physical behaviour, is the tropopause. It appears
as a sharp increase in the stability of the stratification
(potential temperature strongly increases with height) and is
found at a height of about 8 km at the poles to about 16 km in
the equatorial regions. Below the tropopause, in the troposphere,
the air is well mixed, except for water vapour. Apart from
clouds tropospheric air is rather transparent to solar radiation
but almost opaque to terrestrial radiation. The major absorbing
gases in the troposphere are water vapour and carbon dioxide.
Above the troposphere the most important feature is the ozone
layer, extending from about 15 to 40 km height. In this layer
a substantial amount of solar radiation is absorbed and air

1

temperature is high (well over 240 K, while air temperature at
the tropopause level is about 220 K).

To investigate the effect of different atmospheric
absorptivity for solar and terrestrial radiation, we employ a
'two-layer' model as sketched in Figure 1.1. Atmospheric and
surface temperature is denoted by T_a and T_s, respectively.
The solar constant S measures the intensity of the solar beam
when it reaches the earth. It is the energy flux through a
surface of unit area perpendicular to the beam. A current
value of S is 1365 W/m^2. Since the area of interception by
the earth is one quarter of the total area of the earth, the
mean amount of solar energy received at the 'top' of the
atmosphere equals $S/4 = S'$.

Of the incident solar radiation a fraction A_a is absorbed in
the atmosphere and a fraction A_s heats the surface of the earth.
So the planetary albedo is $1-A_a-A_s$. The surface is assumed to
radiate as a black body. Denoting the absorptivity (and
emissivity) of the atmosphere by ε, an amount of $(1-\varepsilon)\sigma T_s^4$ is
emitted to space directly (σ is the Stefan-Boltzmann constant).
The atmosphere radiates an amount of $\varepsilon\sigma T_a^4$ both up and downwards.

In the absence of a dynamical heat flux from surface to
atmosphere, the long-term global energy balance is found by
setting the energy fluxes through the surface and at the top
of the atmosphere equal to zero. This yields

(1.1.1) $\qquad S'A_s + \varepsilon\sigma T_a^4 - \sigma T_s^4 = 0$, and

(1.1.2) $\qquad S'(A_a+A_s) - (1-\varepsilon)\sigma T_s^4 - \varepsilon\sigma T_a^4 = 0$.

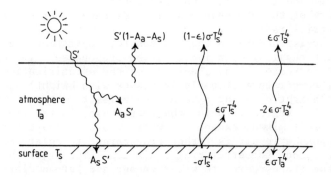

Figure 1.1. A simplified model of the earth's
radiation budget.

From these equations we immediately find that

(1.1.3) $T_s = \{S'(2A_s+A_a)/(2-\varepsilon)\sigma\}^{1/4}$, and

(1.1.4) $T_a = \{S'(A_s+A_a/\varepsilon)/(2-\varepsilon)\sigma\}^{1/4}$.

Equation (1.1.3) shows that surface temperature increases when
the emissivity of the atmosphere increases. The same applies to
atmospheric temperature when A_a is small. It can also be seen
that the surface is warmer than the atmosphere if $A_s/A_a > (1-\varepsilon)/\varepsilon$.
    Typical values of the absorption of solar radiation are $A_a=0.2$
and $A_s=0.5$. Figure 1.2 shows how mean air and surface temperature
then depend on ε. Both T_a and T_s increase with increasing
atmospheric emissivity. The mean vertical temperature gradient,
defined as $\gamma = (T_s-T_a)/Z$, where Z is the height of the centre of
gravity of atmospheric mass, also increases with ε. Since Z is
close to 5 km, and the atmospheric stratification becomes
unstable when $\gamma = 6$ K/km (for typical moisture conditions),
upward energy transport by convection occurs when $T_s-T_a > 30$ K.
According to Figure 1.2 this is the case when $\varepsilon > 0.78$. For the
present atmospheric composition $\varepsilon \simeq 0.9$, so the convective

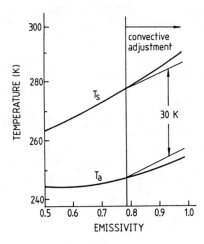

Figure 1.2. Mean air temperature (T_a) and surface
temperature (T_s) as a function of atmospheric emissivity
ε. For $\varepsilon > 0.78$, convective adjustment keeps T_s-T_a
constant.

energy flux plays an important role. It lowers surface
temperature and raises atmospheric temperature. In fact the
upward heat flux is so large that γ remains constant for larger
values of ε. This feature is termed convective adjustment.

The model described above gives a reasonable description of
the nature of the global mean energy balance, but local
differences are large. Convective energy fluxes mainly occur in
lower and middle latitudes, and strongly depend on the particular
season. Also, A_a, A_s and ε show large variations in space and
time. Absorption of solar radiation depends on cloudiness,
aerosol concentration and surface albedo; ε depends in particular
on cloudiness and water-vapour concentration.

An even simpler model of the global radiation balance can be
formulated if we assume that the vertical temperature profile is
invariant under climatic change. In that case the atmospheric
state can be represented by the temperature at one single level.
Surface temperature is most suitable for this purpose because it
allows to take into account the temperature – surface albedo
feedback. Denoting the planetary albedo by α, the radiation
balance reads

$$(1.1.5) \qquad S'(1-\alpha) = \tau\sigma T_s^4 \ .$$

Here τ is the effective transmissivity of the atmosphere for
radiation emitted by the surface. Inserting T_s = 287 K and
α = 0.3 yields τ = 0.62.

Since the range of interest over which T_s varies is rather
small, it is appropriate to linearize (1.1.5) around the present
temperature, i.e.

$$(1.1.6) \qquad S'(1-\alpha) = \tau\sigma T_s^4 + 4\tau\sigma T_s^3 T_s' = a + b T_s' \ .$$

Values of a and b can be found by setting T_s = 287 K and τ = 0.62.
We find a = 238.5 W/m^2 and b = 3.32 W/m^2K.

Of particular interest is the value of b, because it governs
the sensitivity of T_s to changes in the amount of absorbed solar
radiation. Differentiating (1.1.6) with respect to S' shows that

$$(1.1.7) \qquad dT_s/dS' = (1-\alpha)/b \ .$$

So for a larger value of b the climate is less sensitive to
changes in the solar constant. For the value of b given above it
follows that a 1 % drop in the solar constant leads to a
temperature decrease of 0.72 K.

In reality many processes in the atmosphere affect the way in
which outgoing radiation depends on temperature. The most
important one probably is the variable water-vapour content.
Increasing air temperature leads to a higher water-vapour
concentration, thereby increasing the counterradiation from air
to surface. This implies that a larger increase in surface
temperature is needed to restore a perturbed energy balance. Or,
in other words, climate sensitivity is enhanced. In terms of
(1.1.7), this enhancement should be reflected by a lower value
of b.

One way to find a value of b is to employ zonal climatology.
Figure 1.3 shows a plot of outgoing infrared radiation as
measured by satellite versus surface temperature. Each symbol
represents mean values over a latitude belt of 10° width.
Apparently, a clear relation between surface temperature and
outgoing radiation exists, and in fact the data suggest that
b = 2 W/m^2K. Now a 1 % reduction in S' would lead to a
temperature drop of 1.2 K !

Some objections can be made against using zonal climatology.
The water-vapour feedback is automatically included, which is
desirable. However, the latitudinal variation of cloudiness
gives a spurious contribution to the relation between surface
temperature and outgoing long-wave flux, because the distribution
of clouds is more directly linked to the atmospheric circulation
than to surface temperature. Other factors (land-sea distribution,
latitudinal differences in atmospheric stratification) may lead

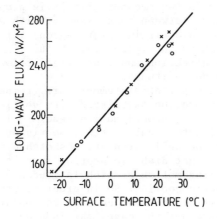

Figure 1.3. A plot of surface temperature vs net
outgoing long-wave radiation. Each symbol represents a
10° wide latitude belt. Crosses refer to the southern
hemisphere, circles to the northern hemisphere.
From Oerlemans and Van den Dool (1978).

to a similar bias. Nevertheless, in case of small perturbations
in the external forcing of the climate system, the procedure
discussed above provides a simple first estimate of climate
sensitivity.

A more detailed discussion on the radiation budget of the
atmosphere can for instance be found in Sellers (1965) and
Lockwood (1979). An instructive model of radiative equilibrium
in the atmosphere is given in Houghton (1977), chapter 2.
For a comprehensive treatment of radiative processes the reader
is referred to Paltridge and Platt (1976).

1.2   The zonal mean state

The amount of solar energy absorbed at lower latitudes is much
larger than the amount absorbed at higher latitudes. A very large
equator-to-pole temperature gradient would be needed to maintain
a state of radiative equilibrium. However, any temperature
gradient involves differences in air density and thus leads to
pressure gradients. So acceleration of air parcels is inevitable
and heat transfer by motions results, implying that a state of
pure radiative equilibrium is not possible. To see this more
clearly, let us have a look at a simple atmospheric circulation
system in a meridional plane.
    In case of hydrostatic balance (which is present for all
atmospheric motions with a horizontal scale larger than about
10 km), the distance between two pressure levels is proportional
to the mean air temperature between those levels. With warm air
in equatorial regions and cold air in polar regions this leads
to a pressure field as shown in Figure 1.4. The air is
accelerated in clockwise direction and thus rises at low
latitudes and sinks at high latitudes.
    Since in the troposphere potential temperature (temperature
corrected for adiabatic expansion associated with vertical
motion) increases with height, upward motion leads to local
cooling and downward motion to local heating. So rising motion
in the upward branch of the cell circulation sketched in
Figure 1.4 counteracts the net diabatic heating in that region.
At the same time the sinking motion at high latitudes
counteracts the net diabatic cooling. So in order to achieve
an energy balance, equatorial air must be warm and polar air
must be cold. If this were not the case, the induced
circulation would tend to increase the pole-equator temperature
difference.
    Atmospheric motion is subject to friction, and if the flow
pattern is to be in equilibrium the acceleration due to the
pressure gradient force must equal the frictional force. In
the absence of other forces, air thus always flows from high

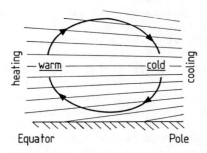

Figure 1.4. A simple cell circulation in the
atmosphere. The thin lines represent isobars.

to low pressure.

In discussing this simple circulation system we have ignored
the fact that the earth is rotating. To see the effect of the
Coriolis force on a meridional cell circulation, we recall that
the force is proportional to the velocity and acts perpendicular
to the velocity vector. Looking downstream, the force is to the
right in the northern hemisphere. The upper level flow in
Figure 1.4 will get a westerly component, and the low-level flow
will come from the east. The resulting type of circulation is
called a Hadley cell.

When forces due to friction and curvature of the air-parcel
trajectory are small (which is usually the case for atmospheric
systems with a horizontal scale larger than 1000 km), the
Coriolis force will balance the pressure gradient force. In
that case a geostrophic balance is said to be present and the
wind blows parallel to the isobars. Large-scale atmospheric
systems, one of which is the Hadley cell, are generally quasi-
geostrophic. Deviations from a pure geostrophic balance are
very small except when the curvature of the isobars is
substantial or friction is large (which is the case near the
surface, in general). Nevertheless, the small deviations from
geostrophic equilibrium account for the conversion of potential
energy into kinetic energy (and vice versa). This is obvious
from the consideration that an air parcel moving parallel to
isobaric levels does not do any work against pressure.

Based on the foregoing discussion we could expect that the
atmospheric circulation can be perfectly organized by Hadley
cells extending from the equator to the poles. In reality this
is not observed. The actual Hadley cells, which vary continuously
in strength and size, have their upward branch in the tropical
regions whereas the downward motions occur over the subtropics

(creating very dry conditions). So why are the Hadley cells so
small ? The answer seems to be that they are not efficient
enough in transporting heat polewards. The remaining south-north
temperature gradient is so large that baroclinic instability
occurs: perturbations on the zonal (i.e. west-east) flow grow by
converting potential energy into kinetic energy to form the
well-known midlatitude storms. These storms (depressions) have
a profound influence on the general circulation of the atmosphere
and are very efficient in transporting heat and moisture.

Although this is not the place to deal with baroclinic
instability in some depth, a brief sketch of the mechanism may
be useful. To this end we consider Figure 1.5. The upper part
of the figure shows isotherms (dashed lines), which at the same
time represent isobars because we assume that a hydrostatic
balance exists. We simple associate this with an almost perfectly
geostrophic Hadley circulation, so the flow is zonal. Next we
suppose that for some reason a wavy perturbation is generated
on this flow. In the figure this is indicated by the heavy line
with open arrows. This perturbation will cause a northward flux
of warm air at A and a southward flux of cold air at C. The warm
air enters into a relatively cold environment and will start to
rise. At the same time the cold air starts to sink. In this way
a cell circulation similar to the one depicted in Figure 1.4 is
initiated, but turned by 90°. Associated with the zonal

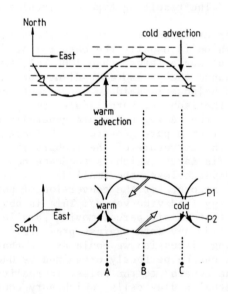

Figure 1.5. Illustration of the mechanism of
baroclinic instability.

temperature gradient created in this way are zonal pressure gradients. Two arbitrary pressure levels (p_1 and p_2) are shown in the lower part of the figure.

In case of instability, the winds generated by the perturbation pressure gradients should enforce the poleward advection of warm air and the equatorward advection of cold air. In the situation sketched in the figure this is not the case. A meridional wind component is generated at B, but not at A and C (because the zonal pressure gradient is zero). So the perturbation cannot grow because the pressure perturbation is precisely in phase with the temperature perturbation.

At this point we should realize that the basic zonal flow changes its speed with height. This is clear from Figure 1.4: a temperature difference between equator and pole implies a difference in spacing of pressure levels. So the horizontal pressure gradient, and therewith the geostrophic zonal wind, must change with height. A direct consequence is that the phase of a wave-like temperature perturbation is forced to change with height. In case of hydrostatic balance, pressure essentially is a vertical integral of temperature, so pressure and temperature perturbations now get out of phase. The result is a shift in the location of the cell circulation. In can be shown that, in case of a sufficiently large south-north temperature gradient, the wave will grow (its kinetic energy increases) when it is tilted westward with height. The wavelength of maximum growth rate is about 5000 km, depending somewhat on the basic state and the latitude at which the perturbation is generated.

As mentioned above, the midlatitude baroclininc waves or storms have a large effect on the general circulation of the atmosphere. They are important both with regard to the energy and momentum balance, restricting the extent of the Hadley cells. Also, precipitation in polar regions is closely connected to the activity of storms: they transport large amounts of water vapour polewards.

Figure 1.6 gives a summary of fluxes that maintain the zonal and annual mean balances of energy, water vapour and momentum. It applies to the northern hemisphere. Fluxes by the meridional overturning, baroclinic waves and ocean currents (in the case of energy) are shown separately. From the figure it is very clear that in the subtropics the Hadley circulation is the major transporting agency, whereas at middle and high latitudes the baroclinic waves take care of the fluxes. The role of the oceans is not insignificant. In particular in the subtropics ocean currents carry a substantial part of the total northward energy flux.

To understand the latitude dependence of the momentum flux it is necessary to have a look at the global momentum balance. Due to surface friction the atmosphere and solid earth exchange momentum. Neglecting the effect of tidal friction, an equilibrium between atmospheric and solid earth momentum exists. The annual

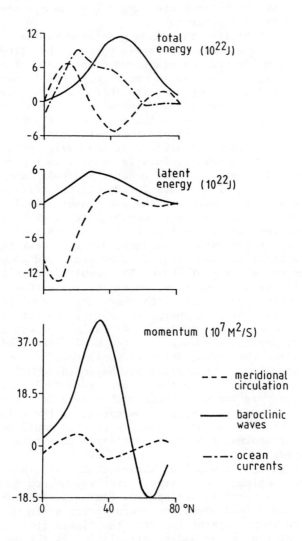

Figure 1.6. Zonal mean annual transports of total energy, latent energy (water vapour) and momentum, in the northern hemisphere. Transport by baroclinic waves refers to transport by all atmospheric motions except overturning in the zonal mean state. Poleward transport is positive.
The figure is based on data from Oort and Rasmusson (1971) and Oort and VonderHaar (1976).

mean total momentum of the atmosphere is fairly constant,
implying that the net momentum exchange with the solid earth is
about zero. In the subtropics surface winds are easterly, so
the atmosphere looses easterly momentum, i.e. gains westerly
momentum. At the same time the atmosphere gives up westerly
momentum to the solid earth at higher latitudes where surface
winds are predominantly from the west. So there must be a steady
poleward flux of westerly momentum, as shown in Figure 1.6.

 We conclude this section by a quick look at the atmospheric
energy cycle. The difference in diabatic heating between polar
and equatorial regions generates available potential energy.
The term available is used here to indicate that not all
potential energy of the atmosphere can be used for conversion
into kinetic energy (it is not possible to bring the centre of
gravity of atmospheric mass down to the surface !). As discussed
above, baroclinic waves draw their energy from the reservoir of
potential energy. According to Figure 1.7, which gives a global
and annual mean balance, most of the kinetic energy of
midlatitude storms is dissipated by internal and surface
friction. A very small part of it is transferred to the zonal
flow through nonlinear interaction. Although the Hadley cells
convert available potential energy into zonal kinetic energy,
on the global scale the conversion is in the opposite direction.
This has to do with the action of the baroclinic waves, which
feed momentum into the zonal flow. The zonal flow therefore
becomes too strong for the existing meridional temperature
gradient. To restore a hydrostatic balance this temperature

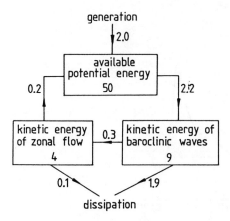

Figure 1.7. The energy cycle of the atmosphere. The
unit for energy reservoirs is 10 000 J/m^2; for energy
conversions 1 W/m^2.

gradient, and therewith the available potential energy must increase slightly.

From Figure 1.7 we see that the amount of kinetic energy in the atmosphere is considerably smaller than the amount of available potential energy. Also, dissipative heating is two orders of magnitude smaller than the difference in diabatic heating between polar and equatorial regions. This demonstrates that the atmospheric heat engine works at a low efficiency.

A very appealing text on the general circulation of the atmosphere is the one by Lorenz (1967). Holton (1972) gives a basic discussion on the problem of baroclinic instability, while a more comprehensive treatment can be found in Pedlosky (1979).

1.3   Zonal asymmetry

Surface conditions along a latitude circle vary widely, in particular in the midlatitudinal and subpolar regions. This so-called zonal asymmetry is forced by the land-sea distribution. The thermal and radiative properties of the oceans differ substantially from those of the continental surfaces. The surface albedo over water (typical value: 0.1) is generally smaller than the albedo over land (typical value: 0.2). However, this difference is partly cancelled by the fact that clouds (which have a high albedo, ranging from 0.3 to 0.6) appear more frequently over ocean than over land. Differences in planetary albedo are therefore not very large.

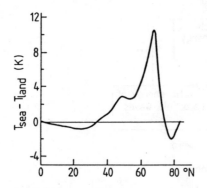

Figure 1.8. Difference in sea-level temperature between oceanic and continental regions in the northen hemisphere.

Another way in which zonal asymmetry can be established is
by poleward energy transport by ocean currents. This tends to
reduce temperatures in the subtropical oceanic regions, while
temperatures in the subpolar oceanic regions increase. It
appears that this effect is important and gives rise to
significant temperature differences between oceanic and
continental regions at high latitudes. Figure 1.8 illustrates
this point. In particular between 60 and 70 °N temperature
over the oceans is much higher. The peak is mainly due to the
North Atlantic Current, which brings a large amount of heat
into the Norwegian Sea.

Atmospheric dynamics also play an important role in creating
zonal asymmetry. Mountain ranges and differences in diabatic
heating rates force quasi-stationary waves in the atmosphere.

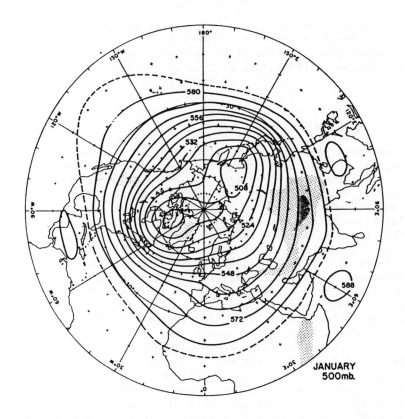

Figure 1.9. Height contours of the 500 mb pressure
level (unit: 10 geopotential meters), for mean
January conditions in the northern hemisphere.
From Palmén and Newton (1969).

Figure 1.9 shows an example. It displays the mean January atmospheric state in the northern hemisphere in terms of height contours of the 500 mb pressure level. This is qualitatively equivalent to the field of isobars at a height of 5 km. Since the wind blows almost parallel to the isobars, there will be regions where enhanced advection of cold polar air occurs frequently (in the case shown, this applies for example to the eastern part of the United States and Canada). Other regions will have an increase in advection of warm air.

The phase and amplitude of these planetary waves depends strongly on the forcing at the surface and the basic zonal flow in which they appear. Therefore the wave pattern is highly dependent on season. The presence of large continental ice sheets will of course have a strong effect on the location and amplitude of the planetary waves in the atmosphere, in particular when they are located in the westerlies. We return to this later.

A study of zonal asymmetry with an energy-balance climate model can be found in Oerlemans (1980). A few recent papers on planetary waves in the atmosphere are Lau (1979), Opsteegh and Van den Dool (1980), and Hoskins and Karoly (1981).

1.4   The oceans

The oceans play a very important role in the global energy budget, because large amounts of heat can be stored in them. To raise the mean ocean temperature by 1 K, 5.8×10^{24} Joules are needed, while a similar temperature increase of the atmosphere requires only 5.7×10^{21} Joules. It would not take too long to establish such a temperature increase of ocean and atmosphere: the infrared emission to space should cease during 348 days only. This will not happen, of course, but it illustrates that the climate system is in a delicate balance between large fluxes of solar and terrestrial radiation. This is even more true when we realize that on time scales shorter than 100 yr the inertia of the deep ocean can hardly be used to damp imbalances in the radiation balance. This is a consequence of the small rate of vertical exchange in the oceans.

As discussed earlier, ocean currents carry a substantial amount of heat polewards and thus have a pronounced effect on climate in the polar regions. But even the global mean temperature is likely to be much higher when ocean currents can reach the poles. At present this is not possible in the Antarctic region and hardly possible in the Arctic region (only two narrow trenches in the North Atlantic Ocean exist that permit some exchange of water between the Arctic Sea and the other oceans), but in the earth's history this situation has been exception rather than rule. So although at present we are

in an interglacial period within the Pleistocene, the earth goes
through a cold phase of longer duration: the presence of snow and
ice has dominated the polar climate for the last millions of
years. Ice sheets on the Antarctic continent probably formed
already 30 or 40 million years ago. Before that time the location
of the continents was such that ocean currents could bring large
amounts of heat all the way down to the poles. Consequently, the
high-latitude albedo was much smaller and therefore the global
climate warmer.

Nevertheless, the present poleward heat flux in the oceans is
not unimportant with regard to conditions at high latitudes. One
way to show this is to take a simple energy-balance climate model
(which we will discuss in detail later), and switch off the
oceanic heat flux. Figure 1.10 shows what would happen to the
annual mean surface temperature in the Northern Hemisphere. The
tropics and subtropics would become warmer, and the high-latitude
regions cooler. At 65 °N the temperature drop would even be 10 K !
The temperature decrease averaged over the hemisphere, which is
due to the dependence of albedo on temperature, would only be 1.4 K.

So heat transport by ocean currents is important, and we expect
any change in the oceanic circulation to have an effect on climate.
In view of this a few remarks on the nature of the large-scale
circulation of the ocean are in order.

The ocean circulation is driven in two ways. Firstly, by the
wind stress action on the water surface, and secondly by density
gradients associated with differences in water temperature and
salinity. Variations in salinity are due to differences in the
water balance at the surface (precipitation minus evaporation) and

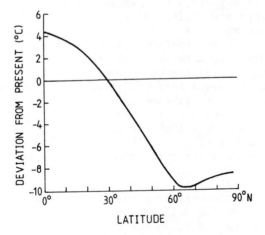

Figure 1.10. Decrease in zonal mean temperature if
the heat flux due to ocean currents would cease. From
Oerlemans (1980b).

river run-off. Formation of sea ice and ice growth at the bottom of ice shelves also leads to the formation of salinity gradients.

The major processes driving the so-called thermohaline circulation in the ocean are: (i) diabatic heating of surface water between roughly 40 °S and 50 °N and cooling outside this belt, and (ii) a negative surface water balance in the subtropical regions due to very high rates of evaporation. The cold water produced in the subpolar regions sinks and gradually flows equatorwards, so there must be some upward motion in the midlatitude and (sub) tropical regions. In the subtropics upward motion is somewhat suppressed by the density increase associated with (ii).

Diabatic heating takes place at the ocean surface and thus stabilizes the stratification. Turbulence generated by surface waves tries to mix the heat downwards, but has to 'struggle' against the stable stratification. In combination with the small upward motion this tends to produce a level below which temperature is fairly constant. The depth of this level varies between a few hundreds of meters and about 1 km.

Diabatic cooling, on the other hand, tends to destabilize the stratification. Convection is initiated, and consequently the high-latitude ocean is relatively well mixed. So a marked asymmetry between regions of cooling and heating shows up. Such an asymmetry also exists in the atmosphere: intense surface cooling at high latitudes tends to stabilize the stratification whereas surface heating in the tropics leads to deep convection.

In spite of this similarity zonal mean overturning in the oceans is much weaker than in the atmosphere. This is due to the fact that the thermal expansion coefficient of a liquid is much smaller than that of a gas. So the density differences in the oceans, and therefore the horizontal pressure gradients that drive the overturning, are small. A typical value of horizontal velocity in the thermohaline ocean circulation is 0.1 m/s.

Surface winds are much more powerful in generating ocean currents. In the equatorial regions a complex system of currents prevails, with many seasonal features (associated with the monsoon circulations in the atmosphere). With regard to the mid- and high-latitude climate, the big gyres centered around 30 ° latitude are most important. They are driven by the easterly trade winds in the subtropics and the westerlies in the midlatitudes, and account for the large poleward heat flux discussed earlier.

A striking feature of these wind-driven current systems is the existence of very strong currents along the western boundaries. Well-known examples are the Gulf Stream along Florida an the Kuro Shio Current along Japan. In the core of these currents velocities are over 1 m/s. Currents in the eastern parts of the oceans are much weaker, so the centres of the gyres are displaced in westward direction. The reason for this shift is the variation of the Coriolis parameter f (= $2\mu \sin\phi$; ϕ is latitude and μ the

angular velocity of the earth), which can be understood as follows.
    A typical subtropical ocean gyre in the Norther Hemisphere, in
terms of streamlines, is sketched in Figure 1.11. Meridional shear
of the zonal surface wind (arrows in the Figure) generate
anticyclonic voticity. In case of a steady state, this has to be
balanced by advection of planetary vorticity and by friction
which dissipates the vorticity (at a rate D). Denoting north- and
eastward velocity components by v and u, and the wind-stress field
by τ, the vorticity budget of the flow reads (anticyclonic
vorticity has negative sign)

$$(1.5.1) \qquad \text{curl}\tau - D[\frac{\partial v}{\partial x} - \frac{\partial u}{\partial y}] - v\frac{\partial f}{\partial y} = 0 \ .$$

In the western part of the gyre the advection of planetary
vorticity ($-v\partial f/\partial y$) is negative, and since curlτ is also negative
the dissipation of anticyclonic vorticity must be large. This is
only the case when the vorticity itself is large, implying that
$|\partial v/\partial x|$ is large (we consider the region where $\partial u/\partial y$ is small).
In the eastern part of the gyre advection of planetary vorticity
and vorticity generation by the wind stress have opposite signs.
Consequently, the amount of dissipation needed to close the
vorticity balance is much smaller now, and so is $|\partial v/\partial x|$.
    The picture sketched above is an extremely simplified one. The
complex geometry of coastlines has a large effect on the pattern
of mass flow in the oceans. In fact the gyres discussed above
exist because continents are present. Only in the southern
hemisphere a strong current is found that occupies a full latitude
belt: the Antarctic Circumpolar Current.

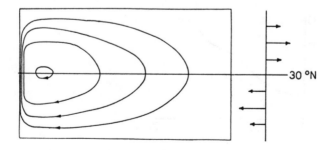

Figure 1.11. An idealized subtropical ocean gyre in
the Northern Hemisphere. Arrows denote surface wind.
From Stommel (1948).

An introductory text in physical oceanography in which the
climatic aspects of the ocean circulation are discussed is
Neumann and Pierson (1966). A more advanced treatment of ocean
dynamics can be found in Pedlosky (1979).

1.5   The seasonal cycle

The climate of the earth is characterized by a strong annual
cycle in most regions. Due to the fact that the earth's rotation
axis is tilted, the insolation at the top of the atmosphere
varies through the year. The effect increases with latitude,
as illustrated in Figure 1.12.
    The upper curve shows the annual range of daily mean
insolation as a function of latitude, and the lower curve gives
the annual temperature range (difference between warmest and
coldest month) that apparently results. These temperatures are
zonal mean values, locally the annual range may be much larger
(up to twice the values shown).
    Although in the middle and higher latitudes the difference
between winter and summer is quite impressive, it is still small
compared to what a calculation of equilibrium climatic states

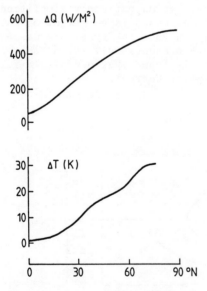

Figure 1.12. Annual range in insolation (from Sellers,
1965) and in zonal mean sea-level temperature (from
Oort and Rasmusson, 1971), as a function of latitude
in the northern hemisphere.

for summer and winter insolation would predict. The reason of course is that the so-called storage term is large. The oceans take up a lot of energy in summer, and give it up again in winter. In view of this it is not surprising that the annual temperature cycle in the northern hemisphere (60 % ocean) is larger than that in the southern hemisphere (80 % ocean).

The effective heat capacity over land is at least one order of magnitude smaller than that over sea (except when large amounts of snow and ice are available for melting), so at a specific location on land the annual insolation cycle will tend to sharpen the contrast between summer and winter, whereas advection of heat by winds will have a damping effect. A characteristic time scale for radiative processes in the troposphere is about 5 days (e.g. Houghton, 1977). With a typical wind speed of 10 m/s for lower tropospheric flow, it follows that advective and radiative processes are equally important when the distance between the ocean and the location on land in consideration is about 5000 km. So in the continental interiors the local radiation balance is able to force a large annual cycle in temperature.

The evolution of glaciers and ice sheets is affected very strongly by the presence of a seasonal cycle. Since, on a monthly basis say, melting rates can be much larger than precipitation rates, summer temperature is a very important factor. In one warm summer month the snow accumulation of an entire year can easily melt and run off. We can be rather sure that if there would be no seasonal cycle, a much larger part of the earth would be covered by glaciers and ice sheets.

Many features in the climate system not mentioned here result from the annual cycle in insolation. A general discussion can be found in any textbook on descriptive climatology, for instance in Lockwood (1979).

2. HOW ICE SHEETS AFFECT CLIMATE

Without any doubt, ice sheets affect climate in many ways.
Here we only consider processes that act on a large scale and
therefore have an impact on the global climate system. More
specifically, we will discuss how the growth and decay of ice
sheets perturbs the global energy budget, how the albedo feedback makes the climate system more sensitive and may lead to the
existence of more than one equilibrium state for given external
forcing, and how the dynamics of ocean and atmosphere may be
affected by the presence of large ice sheets.

2.1 Ice sheets and the global energy budget

To melt a continental ice sheet large amounts of energy are
needed. When an ice sheet is being built up precipitation
reaches the surface in the form of snow, and the atmosphere
does not have to melt it (the energy balance of the atmosphere
thus increases!). Ice sheets in a very cold environment, like
the present Antarctic Ice Sheet, are hardly subject to melting,
so the nearby ocean has to account for this. So the amounts of
energy associated with ice-sheet growth or decay affect the
energy balance of the atmosphere and oceans, depending on the
local conditions. To get a feeling for orders of magnitude we
consider Table 2.1.
    First of all we find that the amount of energy involved in
depositing and melting of an ice volume typical for the
pleistocene glacial-interglacials oscillations is just as large
as the amount of energy required to warm or cool the ocean by
a few degrees K. Although we still do not have a complete
picture of how deep ocean temperatures change during a glacial
period, it is obvious that, in considering differences in the
global energy budget for glacial and interglacial conditions,
the ocean temperature is not less important than the water-ice
phase change.
    It is sometimes argued that deglaciations on a time scale of
5000 yr or so are hardly possible because the energy for
melting would not be available. From Table 2.1 it can be seen
that this is difficult to accept.

A simple calculation shows that 0.085% of the absorbed solar
radiation during 5000 yr is sufficient to melt the ice-age
ice sheets. When at the same the oceans warm by about 4 K,
this figure would be about 0.2%. So it seems that rapid
deglaciation does not upset the global energy balance.
Locally, melting of an ice sheet strongly affects the energy
balance, of course, but atmospheric motions and ocean currents
will easily counterbalance the loss of energy by enhanced
advection of heat. Whether ice sheets melt or not is
determined by the energy balance of the ice surface, i.e. by
local conditions like insolation and surface elevation.
Once conditions are favourable to melt ice, the resulting
perturbation in the local energy balance will in most cases
create both vertical and horizontal temperature gradients,
which initiate increased counter radiation and advection of
heat, respectively. This provides the link with the large-scale
energy balance.

As another example we suppose that an Antarctic ice surge
takes place and brings an amount of one tenth of the present
ice volume in the oceans. A simple calculation shows that, if
this amount of ice melts within 100 yr, the mean energy balance
of the climate system is perturbed by about 0.5 W/m^2.
Recalling that $dI/dT = 2$ $W/(m^2K)$ (I is outgoing infrared
radiation), we see that the resulting temperature drop (averaged

total precipitation	$5.1x10^{14}$	m^3/yr
associated release of latent heat	$1.3x10^{24}$	J/yr
absorbed solar radiation	$3.8x10^{24}$	J/yr
present ice volume	$3.0x10^{16}$	m^3
energy needed to melt it	$9.3x10^{24}$	J
ice-age ice volume	$8.1x10^{16}$	m^3
return to interglacial requires	$1.6x10^{25}$	J
energy needed to warm the ocean by 1 K	$5.8x10^{24}$	J
energy needed to warm the atmosphere by 1 K	$5.1x10^{21}$	J

Table 2.1. Some energy data on the global climate
            system.

over the globe) would only be about 0.25 K. However,
accounting for an associated increase in reflectivity for
solar radiation would lead to a larger temperature drop.
This point will be discussed more extensively in later sections.

2.2   The temperature – ice – albedo feedback

Before discussing the nature of the temperature – ice – albedo
feedback the term albedo should be described properly. We
define the albedo as the ratio of the amount of solar radiation
reflected from a surface to the total amount incident on it.
This surface can be fictive, however: the 'top' of the
atmosphere, for example.
     It is important to make a distinction between surface albedo,
clear-sky albedo and planetary albedo. The planetary albedo is
the albedo as measured at the top of the atmosphere (as seen
from satellites, say). So in studying the vertically-integrated
energy balance of the climate system, the planetary albedo is
the relevant quantity. The clear-sky albedo is the planetary
albedo in the absence of clouds. It is always larger than the
albedo of the underlying surface (the surface albedo), because
part of the incoming solar radiation is reflected by the
atmosphere.
     If we denote cloudiness by N $(0 \leqslant N \leqslant 1)$, cloud albedo by
α_{cl} and clear-sky albedo by α_{cs}, we can write for the planetary
albedo α_p:

(2.2.1) $\alpha_p = \alpha_{cl} + \alpha_{cs} (1 - N)$.

Since clouds generally have a large albedo (varying from about
0.35 at the equator to 0.65 at the poles as a result of
different solar elevation), they contribute substantially to the
mean planetary albedo of the climate system. A few typical
values of the clear-sky albedo are: over sea: 0.13; over
land: 0.22; over ice: 0.57; over snow: 0.62 .
     To understand the nature of the temperature – ice – albedo
feedback, we consider the energy balance of some part of the
climate system:

(2.2.2) $Q [1 - \alpha(T)] = a + bT + F$

The left-hand side represents the amount of absorbed solar
radiation, where the planetary albedo α depends on the mean
surface temperature T.

At the right-hand side we have the outgoing terrestrial radiation in the simple linearized form discussed earlier (a + bT), and the net flux of energy F through the boundaries of the region considered. In a first approximation we write for this term

(2.2.3) $F = c\,(T_{out} - T),$

where T_{out} is the temperature outside the region to which the energy balance applies, and c is a positive constant.
    The energy balance can then be rewritten as

(2.2.4) $Q\,[1 - \alpha(T)] = a^* + b^*\,T,$

        with $a^* = a + cT_{out}$, $b^* = b - c$.

    Expanding ice cover leads to an increasing surface albedo and, without any change in cloud cover, to an increasing planetary albedo. We model this by a piecewise linear function:

(2.2.5)   if $T \leqslant T_0$: $\alpha = \alpha_0$,

                if $T \geqslant T_1$: $\alpha = \alpha_1$,

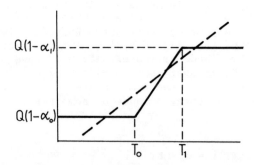

Figure 2.1. Absorbed solar radiation (solid line) and energy loss by outgoing terrestrial radiation and dynamical energy fluxes (broken line).

$$\text{if } T_0 < T < T_1 : \quad \alpha = \alpha_0 + \frac{T - T_0}{T_1 - T_0} (\alpha_1 - \alpha_0).$$

So the amount of absorbed solar radiation depends on temperature as shown in Figure 2.1. The albedo values α_0 and α_1 refer to conditions in which the region considered is totally ice-covered and totally ice-free, respectively.

The broken line in the figure represents the energy loss $a^* + b^* T$. Equilibrium states are thus found as the intersections of solid and broken lines. Obviously, two possibilities exist, depending on the values of b^* and $\mu = -Q(\alpha_1 - \alpha_0)/(T_1 - T_0)$. If $\mu > b$, that is if the albedo feedback is strong, a range of values for a^* exists for which three equilibria occur. On the other hand, if $\mu < b^*$ there is one equilibrium state for any value of a^*.

From these considerations we may conclude that the typical behaviour of the simple climate model is that of the cusp catastrophe (e.g. Gilmore, 1981). For specific parameters the albedo feedback creates bifurcation in the system (1 equilibrium $\leftrightarrow$ three equilibria). The relevant control parameters apparently are b^*/μ and a^*. The equilibrium surface in T, a^*, b^*/μ - space is shown in Figure 2.2. In contrast to the canonical cusp catastrophe, the surface has 'sharp folds', which is due to the use of a piecewise linear function for $\alpha(T)$. Factors favouring the occurrence of three equilibrium states are: strong albedo feedback (large μ), weak infrared damping (small b) and a weak exchange of energy of the considered region with the environment (small c).

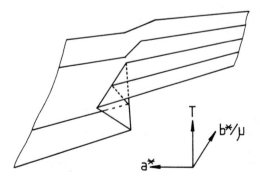

Figure 2.2. Equilibrium surface of the simple energy-balance model with piecewise linear albedo feedback.

The stability of the equilibrium states can be investigated by linearizing the equation

$$(2.2.6) \qquad K \frac{dT}{dt} = Q \, [\, 1 - \alpha(T)] - a^* - b^* T$$

around these states (t is time, K the heat capacity of the system). For equilibria with $T > T_1$ or $T < T_0$ we have $d\alpha/dT = 0$, so the linearized equation then is (a prime denotes deviation from the equilibrium solution):

$$(2.2.7) \qquad \frac{dT'}{dt} = - \frac{b^*}{K} \, T'.$$

Since b^* and K are positive, these states are always stable.

For solutions within the T_0 to T_1 range, we have $d\alpha/dT = \mu$ and the linearized equation becomes

$$(2.2.6) \qquad \frac{dT'}{dt} = \frac{1}{K} \, (\mu - b^*) \, T'.$$

The equilibrium state is thus stable if $b^* > \mu$, but then the total number of equilibria is one (we are outside the cusp of Figure 2.2). In case of three equilibria the intermediate solution is always between T_0 and T_1 ,and unstable because in this situation $b^* < \mu$. We thus arrive at the conclusion that the only unstable equilibrium state in the system is the intermediate solution within the cusp.

At this point we will not yet attempt to find out whether conditions in the climate system (or parts of it) are such that two stable states in the presence of realistic external forcing actually exist. However, in the discussion on ice ages we will pick up this matter again.

2.3 Albedo feedback and external forcing

For very small perturbations of the global energy budget, the response of surface temperature can be assumed to be linear. A quantity characterizing climate sensitivity then is $dT_s/d\Delta$, where Δ is a perturbation of the energy balance. To find this quantity, we consider the perturbation energy balance

$$(2.3.1) \qquad \Delta - Q\alpha' = bT' \quad ,$$

where primes indicate deviations from the reference state.
Writing $\alpha' = -\beta T'$ (β positive), we find

(2.3.2) $$T' = \frac{\Delta}{b - \beta Q} \ ,$$

and the sensitivity parameter thus is

(2.3.3) $$\frac{dT}{d\Delta} = \frac{1}{b - \beta Q} \ .$$

This relation directly shows that climate sensitivity, or
better surface-temperature sensitivity, is enhanced by the
albedo feedback. Curious things happen when $\beta Q > b$; in that
case the solution given by (2.3.2) is unstable and temperature
will either decrease or increase until it comes in a region
where βQ is smaller.
    Another sensitivity parameter which is widely used involves
variations in incoming shortwave radiation (for instant due to
changes in the solar constant). Differentiating the energy
balance with respect to Q yields

$$(1-\alpha) - Q\frac{d\alpha}{dQ} = (1-\alpha) - Q\frac{\partial\alpha}{\partial T}\frac{dT}{dQ} = b\frac{dT}{dQ} \ ,$$

from which it follows that

(2.3.4) $$\frac{dT}{dQ} = \frac{1 - \alpha}{b - \beta Q} \ .$$

So the difference between this sensitivity parameter and the
one defined in (2.3.3) is just $\alpha/(b - \beta Q)$, which is not
surprising.
    In the foregoing analysis dynamic energy fluxes are not taken
into account, and the expressions (2.3.4) and (2.3.3) thus
characterize either global sensitivity (i.e. surface temperature
averaged over the entire earth), or potential local sensitivity.
Here potential indicates sensitivity in case of no change in
energy fluxes. In general, energy fluxes will certainly change
if at some location the energy budget is perturbed (due for
instance to a melting ice sheet, or, on a shorter time scale, a
volcanic eruption). In most cases this change will be such that
the energy perturbation is counteracted. Still (2.3.4) can be
used to quantify the potential of specific regions to destabilize
the climatic state.
    An impression of how dT/dQ varies with latitude is given in

Figure 2.3. The solid curve shows dT/dQ as calculated with an
energy balance climate model. The dashed line shows the result
when energy transports are 'switched on'. According to this
model, the redistribution of energy is so effective that the
sensitivity is about the same everywhere. It is questionable,
however, whether a simple diffusive scheme used in the energy
balance model to calculate the dashed curve is sufficiently
realistic. Anyway, the solid line indicates the presence of a
few regions that are particularly sensitive. One of these is the
latitude belt just south of the northern edge of the northern
hemisphere continents. Because here latitude circles are almost
completely covered by land, the albedo feedback is so strong
that the sensitivity parameter goes to infinity.

  In the winter halfyear snow cover extends far southwards,
and although insolation is smaller, the albedo feedback
still operates. The sensitivity maximum near 35 ^{o}N is due to the
relatively high grounds in this region and the comparatively
low cloudiness [with fixed cloud amount, the albedo feedback
through varying surface conditions is larger when cloudiness is
smaller, see (2.2.1)].

  Sensitivity maxima in the southern hemisphere can be
explained in a similar way. Since here the continents are small,
the extreme values are of less significance.

  At present the northern hemisphere continents do not carry
ice sheets, but from the discussion given above it is obvious
that once they appear, they will enhance climate sensitivity.
This is particularly true because ice sheets can penetrate far
southwards into regions where insolation is large (snow fields
cannot!), and keep the albedo high even in summer.

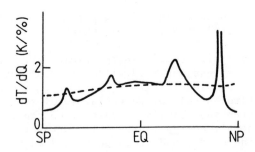

Figure 2.3. Sensitivity of surface temperature to
insolation. The solid line gives dT/dQ as defined in
(2.3.4). The dashed line shows how this curve is
smoothed out when poleward energy transport is taken
into account. From Van den Dool (1980).

2.4 Feedback involving the ocean circulation

As discussed in section 1.5 the general overturning in the
oceans, in which heavier water at high latitudes sinks and
slowly flows equatorward, is partly driven by the temperature
difference of the surface water between polar and equatorial
regions. The presence of ice sheets at high latitudes certainly
affects the ocean temperature. First melting of ice flowing
from continents will keep water temperature low. In the second
place ice sheets tend to cool the local climate through the
ice-albedo feedback. If, for instance, the Antarctic continent
would not carry an ice sheet, the polar and subpolar climate of
the southern hemisphere would certainly be warmer. The
difference in ocean temperature between equatorial and polar
regions will thus depend on whether ice sheets are present.
In this way ice sheets affect the turnover time of the deep-sea
circulation, which is an important parameter for global climate
dynamics [it is of particular interest with regard to the
carbon cycle]. To estimate the possible impact of cryospheric
changes on the deep ocean circulation we will now employ a very
simple model of overturning in the ocean.
   A natural approach to estimate the strength of overturning
in a meridional (north-south, vertical) plane is to use the
circulation theorem from fluid dynamics (e.g. Pedlosky, 1979):

$$(2.4.1) \qquad \frac{dC}{dt} = - \oint \frac{dp}{\rho} + F$$

Here p is pressure, t is time and the circulation C is defined
as

$$C = \oint \vec{v} \cdot d\vec{l} \quad ,$$

in which $\vec{v}$ is the velocity vector and $d\vec{l}$ the infinitesimal
position vector tangent to the (closed) contour. The term F
in (2.4.1) represents all effects but those originating from
the large-scale density gradients. Such effects, like gravity
waves, friction, Coriolis acceleration will generally tend to
destroy circulation in the meridional plane (the Coriolis force
rather turns the circulation plane towards a more zonal
direction).
   In a first approximation it seems reasonable to write

$$(2.4.2) \qquad F = -kC \quad .$$

So we assume that all unknown effects constitute a linear

damping of the circulation.

The circulation, from which a mean velocity can be obtained directly, depends on the choice of contour. To choose a proper one we consider the average meridional density field in the South Atlantic Ocean (Figure 2.4). It is apparent that density increases polewards in the upper layer. With increasing depth, density differences become smaller. Based on this picture a schematic as shown in Figure 2.5 can be used to define the contour. The ocean is split up in two types of water masses, namely, the deep water with constant density ρ_p, and the surface water with density ρ_e at the equator. The warm surface layer generally cools towards the pole.

The integration path is chosen as ABCD. Taking AD, B'C' and BC along isobars the only contribution to the contour integral comes from the sections AB' and CD':

$$\oint \frac{dp}{\rho} = \int_A^{B'} \frac{dp}{\rho} + \int_{C'}^{D} \frac{dp}{\rho} =$$

$$= [\frac{1}{\rho_p} - \frac{1}{\rho_e}] \ (p_{c'} - p_b) \ .$$

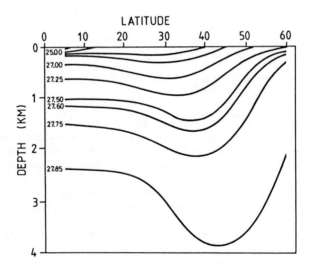

Figure 2.4. Average density field in the South Atlantic Ocean. The quantity shown is $\rho-1000$, where ρ is density in kg/m^3. From Neumann and Pierson (1966).

Applying hydrostatic equilibrium yields

(2.4.3) $\oint \dfrac{dp}{\rho} = gH \ [\dfrac{\rho_e}{\rho_p} - 1 \ ]$,

where H is the distance between the points D and C'.
We can now write the circulation theorem as

(2.4.4) $\dfrac{dC}{dt} = - \ [\dfrac{\rho_e}{\rho_p} - 1 \ ] \ gH - kC$.

    Bryan et al. (1975) estimate the turnover time of the oceans
to be somewhere in the 300 to 1000 yr range. They defined the
turnover time as the ratio of the volume of the deep water to
the rate of the deep-water formation. The generation of
circulation, as formulated by (2.4.4), is directly linked to
the rate of deep-water formation, and we may use the circulation
theorem to estimate for other thermal conditions the turnover
time relative to its present value. An implicit assumption of
course is that the basic circulation system in the world oceans
only changes in strength, not in structure.
    Considering steady states, the ratio of turnover time T to
its present value T_o can be written

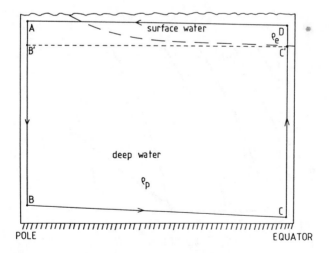

Figure 2.5. The integration path (ABCD) in a meridional
plane used to estimate the magnitude of the deep
ocean circulation.

$$(2.4.5) \qquad \frac{T}{T_o} = \frac{C_o}{C} = \frac{(1-\rho_e/\rho_p)_o}{(1-\rho_e/\rho_p)} \simeq \frac{(\rho_p-\rho_e)_o}{(\rho_p-\rho_e)} \quad .$$

Although the density of sea water depends on both temperature
and salinity, we neglect the effect of salinity gradients
because on the global scale, so density gradients are mainly the
result of temperature differences (salinity gradients are
important with regard to the subtropical/tropical current
systems which, however, do not contribute substantially to the
'renewal' of deep ocean water).

Density of sea water is not a linear function of temperature,
but decreases progressively with increasing temperature for
average salinity conditions. So for a constant temperature
difference the associated density difference becomes larger
when the mean temperature increases. In Figure 2.6 the relative
turnover time T/T_o, as calculated from (2.4.5), is shown as a
function of equatorial (θ_e) and polar (θ_p) water temperature in
the oceanic surface layer. In the calculation the present
situation was assumed to be represented by θ_e = 25°C and
θ_p = -1 °C.

Situations with a constant water-temperature difference lie
on the straight line in the figure. The decrease in turnover
time when the average temperature goes up is obvious. Although
this diagram gives an impression of how T depends on temperature
conditions it does not make clear how ice sheets affect T.

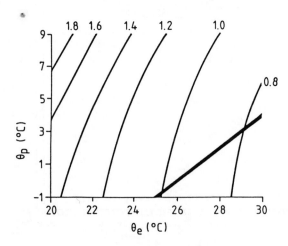

Figure 2.6. Relative turnover time of the deep ocean
circulation as a function of polar and equatorial water
temperature (θ_p and θ_e, respectively).

As noted in the beginning of this section, the presence of ice
in the polar regions will tend to keep polar ocean temperature
low. If water temperature in the polar regions is more tied to
its present value than water temperature in the equatorial
regions, we may infer that due to the presence of polar ice a
general climatic warming will lead to a reduction of the
characteristic turnover time of the world ocean. In case of a
global warming so large that the ice sheets would melt
completely, on the other hand, we expect a much smaller
meridional temperature gradient and then the effects of increase
in mean temperature and decrease in temperature gradient
counteract each other.

So far we did not consider explicitly how the ocean is
actually cooled at high latitudes. There are several mechanisms.
In the first place a large flux of sensible and latent heat from
open water to the atmosphere exists. Cold air flowing from the
high-latitude continents (Antarctica, Siberia, Greenland, Canada)
over sea water can take up large amounts of heat (the heat flux
can easily reach hundreds of Watts per square meter). Sea ice
is very effective in cutting off this energy flux. When sea ice
forms, the heat flux from ocean to atmosphere is reduced by one
or two orders of magnitude. Except in summer when insolation is
large, regions completely covered by sea ice therefore also
become source areas for cold air.

Another mechanism producing dense water is accretion of ice
at the bottom of ice shelves. If the ice is sufficiently cold,
ocean water freezes to the base of the ice shelf. This leads to
formation of rather saline water, which may also help to drive
the overturning in the ocean. It is believed that the Ronne
Ice Shelf in the Weddell Sea is particularly active in this
respect, and forms the so-called Deep Antarctic Bottom Water,
being the most dense water actually present in the world oceans.
Melting at the base of an ice shelf, on the other hand,
produces cold water of low salinity. This water does not sink
before it has been mixed with saline water from outside the
ice-shelf region. In view of these considerations, any
substantial change in the extension and thermal characteristics
of the existing ice shelves is likely to affect the production
of dense water, and therewith the overturning in the oceans.
However, with the present state of knowledge it is hardly
possible to quantify the role of ice shelves.

Melting of icebergs also cools the ocean, but the associated
amount of energy is very small in view of the large area of
ocean generally involved. When conditions are such that icebergs
cannot disperse but steady mixing of melted water with more
saline water can take place, amounts of dense water can be
formed that are of some importance. On a large scale, however,
melting of icebergs plays a minor role in the energy budget.

For further reading, see end of section 1.4.

2.5 Effect of ice sheets on the zonal mean climatic state

Ice sheets occur at high latitudes and reduce the absorption
of solar radiation, so they tend to sharpen the thermal contrast
between equatorial and polar regions. When ice sheets form, the
atmospheric circulation will react to the increasing temperature
gradient and it is likely that the strength of the zonal
circulation will increase. Let us consider in more detail the
effect of glaciation in the northern hemisphere.

In recent years models have been developed that are able to
calculate the zonal mean state of the atmosphere, given the
external forcing (distribution of insolation, optical properties
of the atmosphere). In these models a surface energy balance is
calculated and upward heat fluxes together with divergence of
the radiative flux determine the rate of heating or cooling of
the atmosphere. The albedo feedback is taken into account and
some models even contain a schematic hydrological cycle. The
effects of zonally asymmetric disturbances (mainly the midlatitude storms) are included in a parameterized way. So in
these models poleward fluxes of heat and moisture are
accomplished by two mechanisms: the mean meridional circulation
(overturning) which is calculated explicitly, and the eddies
which are included in parameterized form. In order to calculate
a mean summer state and a mean winter state, storage of heat in
the ocean (in summer) and release of heat from the ocean
(in winter) can be incorporated.

Here we discuss some experiments carried out with such a
model of the zonal mean atmospheric state, originally developed
by Saltzman and Vernekar (1971). An impression of how the zonal
mean climatic state depends on the degree of glaciation can be
obtained by running the model to a steady state for described
and fixed ice conditions. By repeating this for various positions
of the ice line (southern edge of the ice cover) an estimate is
obtained of how climatic variables such as hemispheric
precipitation, strength of the zonal mean circulation, etc.,
vary with the degree of glaciation. A proper way to isolate the
direct effect of ice cover is to keep the distribution of
insolation equal to the present one.

Results of a series of such experiments are shown in
Figures 2.7 and 2.8. The dependence of temperature and
precipitation, averaged over the northern hemisphere, on the
position of the southern edge of the continental ice sheets is
displayed in Figure 2.7. In the experiments on which these
graphs are based, sea-ice extent has been adjusted to be in
reasonable accordance with the extent of the continental ice.
As expected, both hemispheric temperature and precipitation
decrease when the ice line moves southward. The temperature drop
is completely due to the increased albedo of the earth. For an
ice line at 50 °N, the decrease in annual temperature is about
2.5 K, which is somewhat smaller than the drop in global

temperature generally assumed to be present during full glacial
conditions (about 5 K). The reason for the comparatively small
temperature drop lies in the fact that the mean temperature of
the deep ocean water has to be prescribed in the model, and was
set to its present value.

It is not surprising that when temperature decreases
precipitation rates also decrease, at least when hemispheric
mean values are considered. Since fluxes of water vapour from
surface to atmosphere depend on the atmospheric humidity itself,
there is a tendency towards constant relative humidity. Lower
temperature then implies lower absolute humidity, and therefore
a lower precipitation rate. According to the present zonal
climate model, an ice cover up to 50 °N would lead to a 20 %
reduction of hemispheric precipitation.

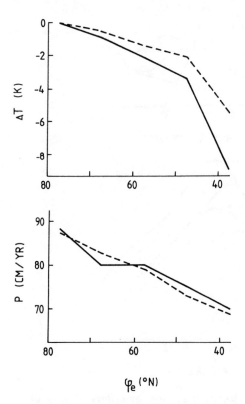

Figure 2.7. Hemispheric mean values of precipitation
rate and temperature drop as a function of the
location of the ice line (in terms of latitude).
Dashed lines apply to the summer, and solid lines to
winter. From Oerlemans and Vernekar (1981).

Figure 2.8 shows the latitudinal distribution of the water balance, as well as the activity of baroclinic waves, for an ice line at 48 $^{\circ}$N. The most striking feature is the enormous increase of the water balance in the midlatitudes and the corresponding decrease in the subtropics. It results from an increase in the poleward flux of water vapour by the baroclinic

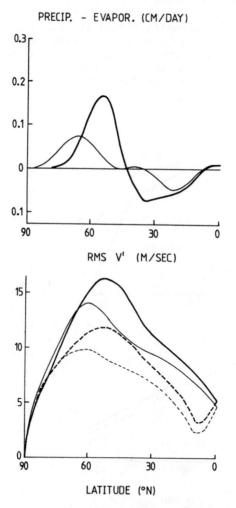

Figure 2.8. Water balance and baroclinic wave activity for a control run (thin lines, reflecting present-day conditions) and an ice-age run (heavy lines). Dashed lines apply to summer, solid lines to winter. The ice-age run has the ice line at 48 $^{\circ}$N. From Oerlemans and Vernekar (1981).

waves, which is a consequence of the increased meridional
temperature gradient near the ice line.

When considering these results, the limitations of zonal
climate models should be kept in mind. Local changes in
precipitation patterns are probably more significant than changes
in zonal mean values. Two-dimensional precipitation patterns for
ice-age conditions have been obtained with the aid of general
circulation models of the atmosphere. However, a comparison of
results from different models shows very large differences in
the distribution of precipitation (Heath, 1979).

2.6 Planetary waves in the atmosphere

As noted in the section on zonal asymmetry, the presence of
large ice sheets in the westerlies may force planetary waves.
There are two ways in which this may happen: mechanically, in
the sense that the ice sheet appears as a barrier to the flow,
and thermodynamically by changing the rate at which the
overlying air is heated or cooled. Here we will deal with the
orographic effect only.

Atmospheric motion on very large (i.e. planetary) scales can
to a first approximation be described by the so-called barotropic
vorticity equation (e.g. Pedlosky, 1979). In a coordinate system
fixed to the earth, this equation reads

$$(2.6.1) \qquad \frac{d}{dt}\left[\frac{\zeta+f}{H}\right] = 0 \ .$$

The quantity in brackets is the so-called potential vorticity,
being the relative vorticity ζ of the horizontal flow plus the
planetary vorticity f divided by the height H of the air column.
So according to (2.6.1) potential vorticity of an air column
is conserved during its motion (friction is not considered).

The relative vorticity of the air column is defined as

$$(2.6.2) \qquad \zeta = \frac{\partial v}{\partial x} - \frac{\partial u}{\partial y} \ ,$$

where u is the (vertical mean) velocity component in the
x-direction (eastward) and v that in the y-direction (northward).
It is easily verified that counter-clockwise motion corresponds
to positive relative vorticity. The planetary vorticity arises
from the component of the earth's rotation vector perpendicular
to the surface of the earth:

$$(2.6.3) \qquad f = 2\ \omega \sin \phi \ .$$

The angular velocity of the earth is denoted by ω, and latitude by φ. So f increases with latitude.

With these definitions and (2.6.1) we are now able to see qualitatively what happens when an air stream encounters a mountain barrier. Let us consider eastward flow over a mountain range that runs from north to south, see Figure 2.9. To keep things simple we assume that upstream of the mountain the velocity is uniform, so ζ = 0. It is obvious that dH/dt is negative at point A, so in order to conserve potential vorticity the relative vorticity should decrease, i.e. become negative. According to (2.2.6) this implies that the motion curves anti-cyclonically. Similarly, when dH/dt becomes positive cyclonic motion will result. The corresponding trajectory is shown in Figure 2.9b.

However, in general the meridional displacement of the air column will be so large that f cannot be considered constant. When the air flows southward f decreases, and conservation of potential vorticity now requires that by the time the air column

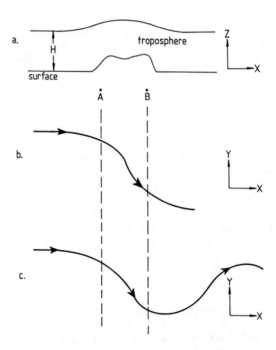

Figure 2.9. Orographic forcing of a Rossby wave in the atmosphere. A cross section is shown in (a). Trajectories in the horizontal plane are given in case of constant f (b), and in case of variable f (c).

reaches point B relative vorticity starts to increase. This brings
the air column far north. Now f becomes large, so ζ should
decrease and anti-cyclonic motion is established. In this way a
mountain range in westerly flow will generate and eastward
propagating Rossby wave, and thus affects the large-scale flow
in the atmosphere over large distances. In the absence of
dissipation, the wave would not damp out.

A full quantitative treatment of forced Rossby waves falls
outside the scope of this book, but it is useful to spend one
more page on free Rossby waves. When H is constant, the vorticity
equation reduces to

(2.6.4) $$\frac{d}{dt} (\zeta+f) = 0 \quad .$$

To carry out a linear analysis we first introduce a stream
function $\psi(x,y)$, so

(2.6.5) $$u = - \partial\psi/\partial y \ , \quad v = \partial\psi/\partial x \ , \quad \zeta = \nabla^2\psi \quad .$$

We consider perturbations on a constant zonal flow U, so
$\psi = -Uy + \psi'$, where ψ' is the perturbation stream function.
Substituting this expression in (2.6.4) and retaining first-order
terms only, the following perturbation vorticity equation
results (using $d/dt = \partial/\partial t + \vec{v}.\nabla$):

(2.6.6) $$\frac{\partial^3\psi'}{\partial t\partial x^2} + U \frac{\partial^3\psi'}{\partial x^3} = - \beta \frac{\partial\psi'}{\partial x} \ ,$$

where $\beta = \partial f/\partial y$, assumed to be constant (the so-called β-plane
approximation). For simplicity $\partial^2\psi'/\partial y^2$ has been set to zero, so
phase and amplitude of the wave only depend on x. It is now
easily verified that (2.6.6) permits solutions of the form
$\psi' = A \sin k(x-ct)$ provided that

(2.6.7) $$c = U - \beta/k^2 = U - \beta L^2/4\pi^2 \quad .$$

Here L is the wavelength.

The reason why this classical dispersion relation for free
waves is derived here is that it shows some of the basic
properties of Rossby waves, particularly the strong dependence
of the phase speed c on the horizontal scale.
Long waves ($L^2 > 4\pi^2 U/\beta$) move westward, while short waves
($L^2 < 4\pi^2 U/\beta$) move eastward. The phase speed of the wave is
determined by the competition between advection of relative
vorticity by the mean flow and advection of planetary vorticity

by the southward perturbation flow. For long waves the latter
dominates.

When forcing (and dissipation) is introduced the phase of
the induced stationary waves depends in a similar way on L as
the phase speed does for free waves. We will not prove this, but ε
it is a general result emerging from planetary-wave models
of different complexity. Long waves (which can only be generated
by a forcing field of very large scale) tend to be shifted in
upstream direction, and short waves in downstream direction.

Summarizing, the message of this analysis is that the phase
(and amplitude, not discussed here) of planetary waves forced by
ice sheets depends strongly on the size of the ice sheets. So
a potentially important but complex interaction mechanism exists,
which up till now has not been studied in any detail.

For further reading on planetary waves, see the section on
zonal asymmetry.

3. MODELLING OF ICE FLOW

The most efficient description of ice flow involves continuum
mechanics. From elementary continuum mechanics to ice-sheet
models for climate studies is a long way, at least if one
desires to go step by step through all required simplifications.
A climate modeler wishing to design a dynamical system that
describes the response of an ice sheet to changes in
environmental conditions may follow another route: by looking
carefully at proxy data on the transient behaviour of ice
sheets, at present-day measurements, and at output from
sophisticated models, a lot of corner cutting can be done.
Familiarity with some basic theory of dynamical systems
combined with physical intuition may lead to efficient model
building. One should not consider this style as inferior.
   Irrespective of the style of modelling, it does no harm to
have some knowledge on continuum mechanics, and on how
conservation of mass, heat and momentum can be formulated.
The first section in this chapter deals with this. Then we go
through a hierarchy of models that describe the shape and
evolution of continental ice sheets.

3.1 Some aspects of continuum mechanics

The purpose of continuum mechanics is to describe how a
material deforms if forces act on it. For elastic materials
(and for forces that are not too large), application of a
finite force leads to a new steady state. If the force is
switched off, the material returns to its initial shape.
Plastic materials, on the other hand, do not.
   Time may play a very important role in deformation. It may
very well be that a rapidly varying force causes elastic
deformation, while a slowly varying force leads to plastic
deformation. So a general description of deformation should
account for the 'memory' of a specific material. This may
be stated as

41

(3.1.1)   deformation $= \displaystyle\int_0^t c(t-t') \; \frac{E(F(t'))}{t'} \; dt'$

F is the force that causes the deformation (at t=0, F=0). In general, E will be a monotonic function of F. For many purposes E can be taken proportional to F. The function $c(t-t')$ represents the memory of the material; it is called the memory or creep function.

   Deformation occurs when the balance of forces acting on some material is disturbed. For a proper treatment of deformation we thus need a tool to describe locally all forces that are present. Two types of force can be distinguished, namely, (i) body forces, acting on an amount of mass, (ii) surface forces, or stresses, acting on a surface. Forces of type (ii) are responsible for, or better, formalize the cause of deformation.

   A useful description of the so-called stress state is obtained by defining a stress tensor. In three-dimensional space, we need three surfaces through some point P to describe all 'independent' surface forces in P. It is convenient to use three surfaces that are perpendicular to each other, and oriented along the axes of a Cartesian coordinate system. For each surface, we have three stress components: the normal stress, acting perpendicular to the surface, and two shear stresses, acting along the surface. So the stress tensor is a second-order tensor τ_{ij} . The subscript j denotes the direction normal to the surface, and the subscript i indicates the direction of the stress.

   It can be shown that a material is only in internal equilibrium (no failure) if the stress tensor is symmetric, i.e. $\tau_{ij} = \tau_{ji}$. So the distribution of stress is in fact given by six independent quantities.

   In many applications the properties of deformation do not depend on hydrostatic pressure. It therefore makes sense to consider the deviation of the stress state from pure hydrostatic equilibrium. Pure hydrostatic equilibrium prevails if $\tau_{ij} = -p \, \delta_{ij}$, where δ_{ij} is the Kronecker delta. To make this more concrete, we split the stress tensor:

(3.1.2) $\begin{bmatrix} \tau_{11} & \tau_{12} & \tau_{13} \\ \tau_{21} & \tau_{22} & \tau_{23} \\ \tau_{31} & \tau_{32} & \tau_{33} \end{bmatrix} =$

$$
\begin{bmatrix}
(2\tau_{11}-\tau_{22}-\tau_{33})/3 & \tau_{12} & \tau_{13} \\
\\
\tau_{21} & (2\tau_{22}-\tau_{11}-\tau_{33})/3 & \tau_{23} \\
\\
\tau_{31} & \tau_{32} & (2\tau_{33}-\tau_{11}-\tau_{22})/3
\end{bmatrix} +
$$

$$
\begin{bmatrix}
(\tau_{11}+\tau_{22}+\tau_{33})/3 & 0 & 0 \\
\\
0 & (\tau_{11}+\tau_{22}+\tau_{33})/3 & 0 \\
\\
0 & 0 & (\tau_{11}+\tau_{22}+\tau_{33})/3
\end{bmatrix} .
$$

Or, in short

(3.1.3) $\qquad \tau_{ij} = \tau'_{ij} + \tau_{kk}\,\delta_{ij}\,/\,3.$

The tensor τ'_{ij} is termed the stress deviator.

Next we consider how deformation can be described. Suppose that two points A and B in three-dimensional space are found at A' and B' after deformation. AA' is determined by the translation vector $\vec{r}$ and BB' by $\vec{r}+d\vec{r}$. So $d\vec{r}$ represents the deformation. In a first-order Taylor series we have (in a Cartesian coordinate system x_i, i=1,2,3):

(3.1.4) $\qquad dr_i = \dfrac{\partial r_i}{\partial x_j}\,dx_j$.

The quantities $\partial r_i/\partial x_j$ form a second-order tensor, which can be split up in a symmetric and anti-symmetric part:

(3.1.5) $\qquad \dfrac{\partial r_i}{\partial x_j} = \dfrac{1}{2}\left[\dfrac{\partial r_i}{\partial x_j} + \dfrac{\partial r_j}{\partial x_i}\right] + \dfrac{1}{2}\left[\dfrac{\partial r_i}{\partial x_j} - \dfrac{\partial r_j}{\partial x_i}\right]$.

The second part represents nothing but a rigid rotation, while the first part describes how the material is deformed. It therefore defines the deformation tensor ε_{ij}. so

(3.1.6) $\qquad \varepsilon_{ij} = \dfrac{1}{2}\left[\dfrac{\partial r_i}{\partial x_j} + \dfrac{\partial r_j}{\partial x_i}\right]$.

The elements with i=j measure stretching, the other elements shearing. Figure 3.1 illustrates this for the two-dimensional case. The typical problem of continuum mechanics is to relate

the deformation tensor to the stress tensor. In order to study
the dynamic behaviour of deforming materials, such a relation
must be added to the conservation equations of mass, momentum
and energy.

A rather simple type of behaviour is that of a linear
elastic isotropic material. In that case deformation and stress
are related by six linear equations of the type

(3.1.7) $$\tau_{ij} = 2\mu\varepsilon_{ij} + \lambda\varepsilon_{ii}$$

The elasticity constants μ and λ tell something about the
'resistance' of a material to external forces. For example,
in a first approximation the effect of an ice load on an
underlying lithospheric plate can be studied with (3.1.7).
The constant μ then measures the rigidity of the plate.
Note that for an incompressible material $\varepsilon_{ii} = 0$, in which case
the second term vanishes.

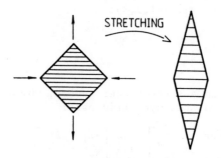

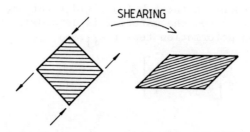

Figure 3.1. Deformation involves both stretching and
shearing.

Equation (3.1.7) describes a simple type of elastic deformation. In the case of plastic deformation, which is more relevant with regard to flow of ice, deformation continues even if the force is constant. A steady state settles down in which particles still move, or in other words, a stationary velocity field is established. Obviously now the rate of deformation should be related to the stress tensor. One of the existing approaches is the following.

As a start we write

(3.1.8) $\dot{\varepsilon}_{ij} = \chi\, \tau'_{ij}$,

where a dot denotes time derivative. χ is not a constant, but may depend on the entire stress state and is generally a function of position. In continuum mechanics the $\dot{\varepsilon}_{ij}$ are called the strain rates. Equation (3.1.8) relates flow to force, and is therefore referred to as a flow law. The degree of sophistication of the flow law, in the form (3.1.8), depends on how detailed χ is modelled. Before this can be done, a few remarks on tensor invariants are in order.

The eigenvalues λ of a second-order tensor σ_{ij} are given by

(3.1.9) $\lambda^3 - I_1 \lambda^2 + I_2 \lambda - I_3 = 0$, where

(3.1.10) $I_1 = \sigma_{ii}$,

(3.1.11) $I_2 = \sigma_{11}^2 + \sigma_{22}^2 + \sigma_{33}^2 + 2(\sigma_{12}^2 + \sigma_{23}^2 + \sigma_{31}^2)$,

(3.1.12) $I_3 = \det(\sigma_{ij})$.

Since the eigenvalues are independent of the orientation of the coordinate system, the quantitites I_1, I_2 and I_3 are invariant under transformation of the coordinate system. Therefore they are called the first, second and third tensor invariant. A general relation between $\dot{\varepsilon}_{ij}$ and τ'_{ij} should also be independent of the choise of coordinate system, so it is natural to assume that

$$\chi = \chi\{I_1(\dot{\varepsilon}), I_2(\dot{\varepsilon}), I_3(\dot{\varepsilon}), I_1(\tau'), I_2(\tau'), I_3(\tau')\} \quad .$$

For applications in this book it is sufficient to consider a flow law involving I_1 and I_2 only. From the definition of the stress deviators τ'_{ij} it follows immediately that $I_1(\tau'_{ij}) = 0$ and consequently $I_1(\dot{\varepsilon}_{ij}) = 0$. For an incompressible material the latter condition is directly satisfied. Since incompressibility will be assumed in all following applications, we only deal with I_2. Writing

$$\dot{\varepsilon} = (I_2(\dot{\varepsilon}_{ij}))^{\frac{1}{2}} \text{ and } \tau = (I_2(\tau'_{ij}))^{\frac{1}{2}},$$

it is postulated that

$$(3.1.13) \qquad \dot{\varepsilon} = A\tau^n$$

Here n is a constant and A generally depends on temperature. The quantities $\dot{\varepsilon}$ and τ are called effective strain rate and effective stress, respectively. From the assumption (3.1.8) and the definition of $\dot{\varepsilon}$ and τ it follows that $\chi = A\tau^{n-1}$, so

$$(3.1.14) \qquad \dot{\varepsilon}_{ij} = A\tau^{n-1}\tau'_{ij}$$

Eq. (3.1.14) is called the generalized flow law. Modelling of the flow in glaciers and ice sheets is based on this law.

For a viscous fluid, n = 1 and A is the dynamic viscosity. In that case the relation between a particular stress component and associated strain rate is linear and independent of the other stresses. For ice the situation is more complex. According to (3.1.14), the strain rate due to a particular stress increases if the other stresses increase (because the effective stress τ increases). Also, the strain rate increases progressively with stress (2 < n < 4, according to laboratory measurements). Later on, more specific examples of how (3.1.14) can be applied to ice flow will be discussed.

3.2 Conservation laws

Modelling of ice sheets and geodynamics should of course be based on proper conservation laws in convenient form. In this section conservation of mass, momentum, and energy are discussed. To close the system, the equations describing these conservation laws should be supplemented by constitutive equations which are typical for the material to be studied.

For example (3.1.7) and (3.1.14) are such equations.
    Conservation of mass is formulated by means of the
continuity equation, which reads

$$(3.2.1) \qquad \frac{\partial \rho}{\partial t} = \frac{\partial}{\partial x_i} (\rho v_i)$$

Density is denoted by ρ, the i-th component of the velocity
vector $\vec{v}$ by v_i. Equation (3.2.1) simply states that density
changes only by convergence of the mass flux $\rho \vec{v}$. For an
incompressible material (3.2.1) reduces to

$$(3.2.2) \qquad \frac{\partial}{\partial x_i} (v_i) = 0 .$$

    The equations of Euler (Newton's law for a continuum)
describe conservation of momentum:

$$(3.2.3) \qquad \rho \frac{dv_i}{dt} = F_i + \frac{\partial \tau_{ij}}{\partial x_j} \quad (i = 1,2,3),$$

where F_i represents the body force. In modelling ice flow,
accelerations are negligible and (3.2.3) reduces to a simple
balance between body forces (gravity in most cases) and stress
gradients.
    For applications in this book it is sufficient to consider
materials that have constant density and constant thermal
conductivity. In that case the energy equation takes the form

$$(3.2.4) \qquad \frac{d\theta}{dt} = \frac{1}{\rho c J} \tau_{ij} \frac{\partial v_j}{\partial x_i} + K \frac{\partial}{\partial x_i} \frac{\partial \theta}{\partial x_i} ,$$

where θ is temperature, c specific heat, J the mechanical
equivalent of heat and K the thermal diffusivity.
So the temperature of a fluid element may change by frictional
heating (first term) and diffusion of internal energy (second
term). Frictional heating occurs when the motion has to
struggle against the stress field, i.e. when it has to do
work.
    The set (3.2.2)-(3.2.4), together with appropriate
constitutive equations, completely describes the state and
motion of a continuum. In many cases an Eulerian description

will be employed, so derivatives 'following the motion' are
written explicitly as

(3.2.5) $$\frac{d}{dt} = \frac{\partial}{\partial t} + v_i \frac{\partial}{\partial x_i} \quad .$$

Readers interested in a more rigorous approach to continuum
mechanics may for instance consult Jaeger (1969).

3.3   Application to ice flow

Deformation of ice, which has a hexagonal crystal structure, is
a complicated process. A useful overview of the mycrophysics of
ice deformation can be found in Paterson (1981). The most
important observation is that if some stress is applied to
ordinary ice, it takes some time before the ice crystals are
arranged into a pattern which is most favourable for deformation.
The macro effect is a deformation rate which slowly increases
with time until a steady state is reached. This type of behaviour
is called creep and a detailed description of ice deformation
should therefore include the use of a creep function, see (3.1.1).
However, in studying ice sheets the time scales of interest are
so large that creep need not be considered.
    As a special and very simple case, we first consider a
situation in which there is only one nonzero stress deviator τ'_{xz}.
Such a situation may for example arise when a slab of ice of
constant thickness and infinite length and width is frozen to
the bottom, while it rests on a surface with constant slope α,
see Figure 3.2. So the analysis can be restricted to two

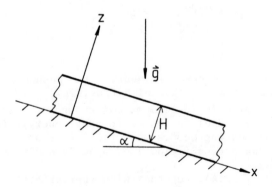

Figure 3.2. Slab of ice on a sloping surface.

dimensions. The only body force acting on the slab is due to the
acceleration of gravity g. Since all derivatives with respect
to x are zero by definition, conservation of momentum reduces to

$$(3.3.1) \qquad - \rho g \sin \alpha = \frac{\partial \tau'_{xz}}{\partial z} \quad , \quad \rho g \cos \alpha = \frac{\partial \tau_{zz}}{\partial z} \quad .$$

At the surface we have $\tau'_{xz} = 0$, so integration from z' = H (ice
thickness) to z' = z yields

$$(3.3.2) \qquad \tau'_{xz}(z) = \rho g (H-z) \sin \alpha \quad .$$

The shear stress at the base, τ_b, thus equals $\rho g H \sin \alpha$, taken
positive when the surface slope is negative (as in Figure 3.2).
In the following, wherever τ_b or $\sin \alpha$ is raised to some power,
its absolute value is used (to keep the notation simple).
    To find the ice velocity we need the flow law. In the absence
of other stress deviators, the effective stress τ just equals
τ'_{xz}, and the resulting relation between strain rate and shear
stress reads

$$(3.3.3) \qquad \dot{\varepsilon}_{xz} = A(\tau'_{xz})^n \quad .$$

From the definition of the deformation tensor (3.1.6) it follows
that $\dot{\varepsilon}_{xz} = (\partial u / \partial z)/2$, so

$$(3.3.4) \qquad \frac{1}{2} \frac{\partial u}{\partial z} = A \, (\rho g \sin \alpha)^n \, (H-z)^n \quad .$$

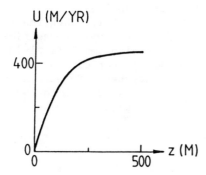

Figure 3.3. Velocity profile calculated from (3.3.5) for
a uniform slab of ice which has a thickness of 500 m.

Another integration with respect to z then yields the velocity profile:

$$(3.3.5) \qquad u(z) = \frac{2A}{n+1} (\rho g \sin \alpha)^n [H^{n+1} - (H-z)^{n+1}] \quad .$$

Figure 3.3 shows a velocity profile according to (3.3.5). Values of the constants used are:

$$n = 3 \ ,$$
$$A = 0.03 \ \text{bar}^{-3} \ \text{yr}^{-1} \ (= 9.51 \times 10^{-25} \ \text{m}^6 \ \text{N}^{-3} \ \text{s}^{-1}) \ ,$$
$$\alpha = 5^o \ ,$$
$$\rho = 900 \ \text{kg m}^{-3} \ ,$$
$$g = 9.8 \ \text{m s}^{-2} \quad .$$

Due to the nonlinear character of the flow law, most of the velocity shear is found in the lower ice layers. In real situations this effect is enhanced by the temperature distribution. As we will discuss in more detail later, the temperature in an ice sheet increases with depth. Since A increases with temperature, strain rates will be larger near the base.

Next we consider a situation in which the bedrock is flat, while the ice surface has a slope $\partial H/\partial x$. The x-axis is along the bedrock, the z-axis points upward. Now the momentum balance reads

$$(3.3.6) \qquad \rho g = \frac{\partial \tau_{zz}}{\partial z} + \frac{\partial \tau_{xz}}{\partial x} \ , \quad 0 = \frac{\partial \tau_{xz}}{\partial z} + \frac{\partial \tau_{xx}}{\partial x} \quad .$$

This case is more complex than the previous one, because stress gradients are not 'decoupled'. Fortunately, when the bedrock is flat and variations in the surface slope are sufficiently small, normal stress deviators are also small (implying that longitudinal strain rates are small). For continental ice sheets such conditions prevail almost everywhere. So we set the normal stresses equal to the hydrostatic pressure, which equals $\rho g (H-z)$. Note that atmospheric pressure is neglected. So we acquire

$$(3.3.7) \qquad \frac{\partial \tau_{xz}}{\partial z} = +\rho g \frac{\partial H}{\partial x} \ ,$$

so

$$(3.3.8) \qquad \tau_{xz} = -\rho g (H-z) \frac{\partial H}{\partial x} \quad .$$

From the flow law we now find

(3.3.9) $\dot{\varepsilon}_{xz} = \frac{1}{2} [\frac{\partial u}{\partial z} + \frac{\partial w}{\partial x}] = A [\rho g(H-z) \frac{\partial H}{\partial x}]^n$,

where w is the upward ice velocity. From (3.3.9) and the
continuity equation ($\partial u/\partial x + \partial w/\partial z = 0$) it is possible to find
the velocity field. By using a streamfunction ψ defined as

(3.3.10) $u = \partial\psi/\partial z$, $w = -\partial\psi/\partial x$,

the continuity equation is immediately satisfied. Substitution
in (3.3.9) yields a Poisson equation:

(3.3.11) $\nabla^2\psi = 2A [\rho g(H-z) \frac{\partial H}{\partial x}]^n$,

where ∇ is the two-dimensional gradient operator. This equation
can be solved by standard numerical procedures when the
boundary conditions on ψ are known. If the ice-accumulation rate
at the surface is M(x), the proper boundary condition is

(3.3.12) $\psi(x,H) - \psi(x,0) = C + \int_{x_o}^{x} M(x')dx'$.

Here C is a constant.

3.4 Perfect plasticity

As discussed in the previous section, in the case of one simple
shearing stress the flow law takes the form

(3.4.1) $\dot{\varepsilon}_{xz} = \frac{\partial u}{\partial z} = A \tau_{xz}^n$,

where the value of n is somewhere between 2 and 4. So for small
stresses the deformation rate is small, while for large stresses
it shows very high values. This suggests that deformation of ice
can be treated to some extent as deformation of a perfectly
plastic material. Here perfectly plastic means that a yield
stress τ_o exists such that for $\tau_{xz} < \tau_o$ the deformation rate
is zero, and for $\tau_{xz} > \tau_o$ it is infinitely large.

So if body forces are present, the ice deforms in such a way that the shear stress equals τ_0 everywhere.

Perfect plasticity can be regarded as an asymptotic case of the flow law (3.4.1). If we set $A' = A \tau_0^n$, (3.4.1) can be written as

$$(3.4.2) \qquad \dot{\varepsilon}_{xz} = A' \, (\tau_{xz}/\tau_0)^n \quad .$$

Taking the limit $n \to \infty$ then shows that $\dot{\varepsilon}_{xz} \to 0$ if $\tau_{xz} < \tau_0$ and $\dot{\varepsilon}_{xz} \to \infty$ if $\tau_{xz} > \tau_0$.

The approximation of perfect plasticity can be very useful. For example, it can be employed to derive quickly a first estimate of the profile of a continental ice sheet. Suppose that the bedrock is flat, and that the ice is frozen to the bottom. Taking the x-axis along the bed and z upward again, (3.3.8) can be used to describe the distribution of τ_{xz}. So at the base we have

$$(3.4.3) \qquad \rho g H \frac{dH}{dx} = \tau_0 \quad .$$

This can be integrated immediately to give

$$(3.4.4) \qquad H = [\frac{2\tau_0}{\rho g} x]^{\frac{1}{2}} \quad .$$

So the profile of a perfectly plastic ice sheet is parabolic. Using $\tau_0 = 0.8$ bar as a typical value gives a profile as shown in Figure 3.4. Note that the profile of a plastic ice sheet

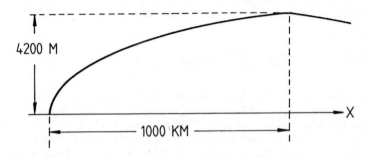

Figure 3.4. Profile of a perfectly plastic ice sheet.

does not depend on the mass balance (the total mass balance
should be positive, of course). Any increase in the mass balance
is compensated for by an equal increase in the ice-mass discharge
to the ice-sheet edge.

The horizontal ice velocity can be calculated directly from
conservation of mass:

$$(3.4.5) \qquad \bar{u}(x) = \frac{1}{H} \int_{\frac{1}{2}L}^{x} M(x')dx' \quad .$$

In (3.4.5) $\bar{u}$ is the vertical mean horizontal ice velocity, M is
the ice-accumulation rate and L is the size of the ice sheet
(distance from edge to edge). If the ice-accumulation rate is
constant this reduces to

$$(3.4.6) \qquad \bar{u}(x) = (x-\tfrac{1}{2}L) \; M/(\sigma x^{\frac{1}{2}}),$$

where we have used the symbol σ for $(2\tau_0/\rho g)^{\frac{1}{2}}$.

At $x = 0$ problems arise. Since H goes to zero when $x \downarrow 0$, $\bar{u}$
must go to infinity to meet the continuity requirement. So at
the edge of the ice sheet the model description fails. In real
situations the ice-sheet edge is subject to melting or calving,
and H does not go to zero in a smooth way. In spite of this
difficulty, the profile shown in Figure 3.4 appears to be a
reasonable first approximation to observed ice-sheet profiles
in regions where no pronounced bedrock topography is present.

To get some idea about orders of magnitude, a calculation
of $\bar{u}$ is useful. For the ice sheet shown in Figure 3.4, with an
ice-accumulation rate of 0.2 m/yr, we find the values shown in
Table 3.1.

x (km)	u (m/yr)
100	−68
200	−43
400	−23
600	−12
800	−5

Table 3.1. Vertical mean horizontal ice velocity for
the ice sheet of Figure 3.4. The ice accumulation rate
is 0.2 m/yr.

3.5 Modelling vertically-integrated ice flow

In many climate simulation studies in which continental ice
sheets play a role, it is not feasible to solve the full set of
stress - strain rate relations in three-dimensional space.
Limited computer power requires efficient ice-sheet models, so
there is a continuous need to look for simplifications that do
not change the properties of an ice-sheet model in an essential
way. The degree of simplification depends on the specific purpose,
of course. If one desires to carry out a paleoclimatic model
study of the Laurentide Ice Sheet, very detailed modelling makes
no sense because boundary conditions (like the ice-accumulation
rate) are poorly known. On the other hand, a study of the
stability of the present grounding line in the Ross Sea requires
careful consideration of all stresses, and simplifications should
only be made on the basis of sound evidence from field
experiments.

When looking at the global characteristics of a large
continental ice sheet, it appears that the horizontal ice
velocity is much larger than the downward ice velocity. Ice flow
can thus be considered as quasi two-dimensional. This does not
mean that vertical motion is unimportant, but it suggests that
a two-dimensional model should be capable of describing the
evolution of an ice sheet. In such a model, processes associated
with vertical motion which affect the horizontal ice flow should
be included in parameterized form.

A two-dimensional model of an ice sheet can be constructed
by integrating the appropriate equations over the vertical
coordinate. First we define a Cartesian coordinate system (x,y,z)
in which the z-axis points upward. The x,y-plane is tangent to
the local bedrock. The velocity vector in this plane is denoted
by $\vec{v}$, and $\nabla = (\partial/\partial x, \partial/\partial y)$. The continuity equation (3.2.2) can
then be integrated as follows:

$$w(H) - w(0) \simeq w(H) \equiv \frac{dH}{dt} = \frac{\partial H}{\partial t} + \vec{v}(H) \cdot \nabla H =$$

$$= -\int_0^H \nabla \cdot \vec{v} \, dz = -\nabla \cdot \int_0^H \vec{v} \, dz + \vec{v}(H) \cdot \nabla H \quad ,$$

so

(3.5.1) $\dfrac{\partial H}{\partial t} = - \nabla \cdot (H\bar{\mathbf{v}})$,

where $\bar{\mathbf{v}}$ is the vertical mean value of $\mathbf{v}$. It has been assumed
that $w(0)$ is small compared to $w(H)$. In most cases $w(0)$ will be
zero; however, if melting at the base occurs $w(0)$ can have
a value of a few mm/yr, which is still small compared to $w(H)$.

For the more general case in which the ice-accumulation rate M
is not zero we have

(3.5.2) $\dfrac{\partial H}{\partial t} = - \nabla \cdot (H\bar{\mathbf{v}}) + M$.

Obviously, if $\bar{\mathbf{v}}$ can be expressed in terms of ice thickness and
bedrock topography, (3.5.2) can be used to study the evolution
of ice sheets.
     At this point it should be noted that (3.5.2) is valid only
when variations in bedrock slope are sufficiently small. If the
curvature of the bedrock is not small, additional ice-mass
divergence occurs because we have taken the x,y-plane tangent to
the local bedrock. Since in general curvature of the bedrock is
much smaller than the ice thickness (this is certainly true for
bedrock topography resolved on a computational grid with a
typical spacing of 10 to 100 km), the effect can be neglected.
     We now turn to the computation of $\bar{\mathbf{v}}$. Two approaches can be
distinguished. The first one was proposed by Nye (1959), and is
based on the observation that most of the velocity shear in an
ice sheet is found in the bottom layer (a consequence of the
high temperatures generally found in this layer). So with regard
to vertically-integrated ice-mass discharge there is not much
difference between internal deformation and basal sliding. It is
thus natural to use a law of sliding of the type

(3.5.3) $\bar{\mathbf{v}} = B|\vec{\tau}_b|^{m-1} \vec{\tau}_b$.

Here $\vec{\tau}_b$ is the basal shear-stress vector, defined as

               $\vec{\tau}_b = (\tau_{xz}, \tau_{yz})$ at the base .

So the vertical mean ice velocity is set proportional to the mth
power of the basal shear stress. B is mostly taken constant, but
can of course be made a function of ice temperature, presence of
basal melt water, etc.
     A second approach is to employ direct scaling of (3.3.3),
leading to

(3.5.4) $\dfrac{\bar{\mathbf{v}}}{H} = A |\vec{\tau}_b|^{n-1} \vec{\tau}_b$.

A vertical mean flow law of this type is more directed towards
the description of ice-mass discharge by internal deformation
throughout the column, without basal sliding. Modellers using
(3.5.4) generally do this in connection with a separate

treatment of basal sliding.

Although there is a principal difference in using either
(3.5.3) or (3.5.4), the practical consequences are not very
large. Many combinations of m and B, and of n and A, can be found
that yield realistic ice-sheet profiles.

Using (3.5.3) or (3.5.4) implies that the effect of normal
stresses is neglected. Since normal stresses are important at
the ice-sheet edge, building a time-dependent model with the
equations discussed above requires a special treatment of the
edge of the model ice sheet.

The last step in obtaining a closed set of equations is to
find an expression for the driving shear stress. The following
equation combines the two special cases discussed in section 3.3
(a slab of ice of uniform thickness on a sloping bedrock, and
an ice mass with surface slope on a flat bedrock) :

$$(3.5.5) \qquad \vec{\tau}_b(x,y) = -\rho g H (\nabla H + \nabla b) \quad .$$

It has been assumed that the bedrock slope is so small that
$\sin \alpha$ can be replaced by $\partial b/\partial x'$ ($b(x',y')$ denotes the bedrock
height with respect to some equipotential level). Note that
primes now refer to a Cartesian coordinate system in which
gravity is down the z'-axis. A complete model can now be
formulated by identifying the (x,y,z) and (x',y',z') systems.
Altogether, this means that the only way in which a bedrock slope
affects the ice flow is by contributing to the local basal shear
stress. No distinction is made between horizontal ice velocity
and ice velocity parallel to the bedrock. This point should be
kept in mind (it becomes important when the release of potential
energy by downward motion is considered).

3.6 Steady-state profiles of an ice sheet

In section 3.4 the profile of an ice sheet was found to be
parabolic in the case that ice can be treated as a perfectly
plastic material. The next step is to replace the plasticity
approximation by a flow law of the type discussed in the previous
section.

If we consider ice flow along a flow line (x-axis), (3.5.3)
can be written as

$$(3.6.1) \qquad u = B \, \tau_b^m \quad .$$

For an ice sheet resting on a flat bottom, the basal shear
stress is

(3.6.2) $\qquad \tau_b = -\rho g H \dfrac{dH}{dx}$.

At the centre of the ice sheet (at the ice divide) the horizontal pressure gradient is zero, implying that there is no flux of ice. In case of a steady state the total accumulation of ice at the surface between the ice divide and an arbitrarily chosen point x must be equal to the mass flux (or volume flux, when density is taken constant). When the accumulation rate M is constant this condition can be expressed as

(3.6.3) $\qquad (x - \tfrac{1}{2}L)\, M = H\, u$.

The ice sheet is assumed to be of size L and symmetric with respect to $x = \tfrac{1}{2}L$.

Inserting the expressions for the vertical mean ice velocity and basal shear stress given above yields

(3.6.4) $\qquad (x - \tfrac{1}{2}L)\, M = B\, H \left(- \rho g H \dfrac{dH}{dx} \right)^m$.

Integrating this equation with respect to x then gives the stationary profile of the ice sheet:

(3.6.5) $\qquad (H/H_o)^{2+\frac{1}{m}} + \left| \dfrac{2x}{L} - 1 \right|^{1+\frac{1}{m}} = 1$.

Here H_o is the ice thickness in the centre given by

(3.6.6) $\qquad H_o^{2+\frac{1}{m}} = \dfrac{2m+1}{m+1} (M/B)^{\frac{1}{m}} \dfrac{1}{\rho g} (\tfrac{1}{2}L)^{1+\frac{1}{m}}$.

In a similar way a profile corresponding to the flow law (3.5.4) can be found. It reads

(3.6.7) $\qquad (H/H_o)^{2+\frac{1}{m}} + \left| \dfrac{2x}{L} - 1 \right|^{1+\frac{1}{m}} = 1$,

with

(3.6.8) $\qquad H_o^{3+\frac{1}{m}} = \dfrac{3m+1}{m+1} (M/A)^{\frac{1}{m}} \dfrac{1}{\rho g} (\tfrac{1}{2}L)^{1+\frac{1}{m}}$.

While the profile of a perfectly plastic ice sheet is independent of the mass balance, the expressions given above show that the profiles based on more realistic flow laws are functions of the mass balance: larger values of M lead to an increase of ice thickness.

In Figure 3.5 the profiles given by (3.6.5) and (3.6.7) are compared to the parabolic profile (3.4.4). The constant m in the flow laws has been set to 3, whereas the constants A and B have been chosen in such a way that the three sheets have the same mean ice thickness. Apparently, the profiles based on the flow laws are generally flatter than the parabolic profile except near the edge of the ice sheet where the ice thickness decreases very rapidly. This implies that the ice velocitiy increases very strongly near the edge. Another difference appears at the centre of the ice sheet. From the figure we see that the assumption of perfect plasticity leads to an ice sheet with a rather sharp ice divide, while the use of a flow law gives a profile in which the halves of the ice sheet fit together in a smooth way. This is of course a more realistic situation.

A comparison of the theoretical profiles derived in this section with actual profiles measured at Antarctica shows that the calculated profiles provide a reasonable description of an ice sheet. It is only very close to the ice-sheet edge or in case of large bedrock irregularities that measured profiles deviate substantially (see for instance Paterson, 1981).

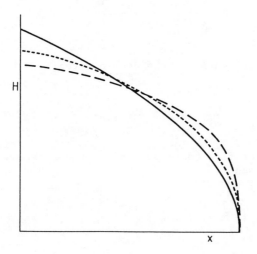

Figure 3.5. Steady-state profiles of a continental ice sheet on a flat bedrock in case of perfect plasticity (———), for the flow law (3.5.3) (····) and for the flow law (3.5.4) (----).

3.7 Normal stresses

The flow law discussed in section 3.5 is based on the assumption
that the shear stress τ_{xz} is the dominant stress component. This
approximation does not hold at the junction of an ice sheet and
an ice shelf (i.e. where the ice starts to float) and in the
ice shelf itself. Also when bedrock topography is very irregular
the use of (3.5.4) becomes discutable. Therefore we now turn to
a discussion of normal stresses.
   The general flow law was formulated as (3.1.14)

$$\dot{\varepsilon}_{ij} = A \, \tau^{n-1} \, \tau'_{ij} \quad , $$

in which τ is the effective stress defined by

(3.7.1) $$2\tau^2 = \tau'^2_{xx} + \tau'^2_{yy} + \tau'^2_{zz} + 2(\tau^2_{xy} + \tau^2_{yz} + \tau^2_{zx}) \quad .$$

In the case of simple shear τ_{xz} and $\dot{\varepsilon}_{xz}$ are the only nonzero
components. The general flow law then reduces to (3.4.1).
In case of uni-axial compression or tension in which τ_{xx} is the
only nonzero stress component, the flow law becomes

(3.7.2) $$\dot{\varepsilon}_{xx} = \frac{2}{9} A \, \tau^3_{xx} \quad , $$

where n has been set equal to 3.
   From (3.7.2) we see that the strain rate produced by a normal
stress is a factor 2/9 smaller than that produced by an equally
large shear stress. This result has been verified experimentally.
   To obtain an expression for the normal stress deviator along
a flow line in an ice sheet or ice shelf, we consider the stress
equilibrium of a column in the ice stream (Figure 3.6).
In the case of two-dimensional flow, the longitudinal stress
deviator is

$$\tau'_{xx} = \frac{1}{2} (\tau_{xx} - \tau_{zz}) \quad . $$

Note that the hydrostatic pressure equals $\frac{1}{2}(\tau_{xx} + \tau_{zz})$.
   When the slope of the bedrock and the ice surface are small,
the net force acting along the x-axis on the column ABCD due to
the difference between the normal stresses on AB and CD is

(3.7.3) $$\frac{\partial}{\partial x} (2H \, \overline{\tau'_{xx}}) \, \delta x \quad . $$

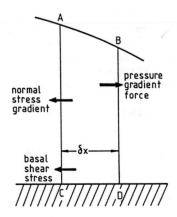

Figure 3.6. Forces acting on a column of ice.

In (3.7.3) the overbar denotes a vertical mean value. The factor 2 appears because τ'_{xx} is defined as the difference between the normal stresses τ_{xx} and τ_{zz} divided by 2.

The surface slope $\partial h/\partial x$ will generally not be zero and will thus give rise to a horizontal pressure gradient. The corresponding total force on the column is

$$(3.7.4) \qquad \int_{b}^{h} \rho g \frac{\partial h}{\partial x} \, \delta x \, dz = \rho g H \frac{\partial h}{\partial x} \, \delta x \quad .$$

Finally, the total force on the ice column due to the shear stress τ_{xz} equals

$$(3.7.5) \qquad \int_{b}^{h} \frac{\partial \tau_{xz}}{\partial z} \, \delta x \, dz = - \tau_{b} \, \delta x \quad ,$$

in which τ_{b} is the basal shear stress. It has been assumed that the shear stress parallel to the ice surface can be neglected.

Equilibrium of forces acting on the column ABCD can now be expressed as (neglecting atmospheric pressure)

$$(3.7.6) \qquad \frac{\partial}{\partial x} (2H \overline{\tau'_{xx}}) = \rho g H \frac{\partial h}{\partial x} + \tau_{b} \quad .$$

This equation enables us to calculate the normal stress deviator along a flow line. Note that if τ_{xx} is set to zero (3.7.6) reduces to (3.5.5).

In ice sheets normal stresses are generally small, except near the ice divide and particularly near the ice-sheet edge. As will be discussed later movement of the ice-sheet edge is closely related to the existence of normal stress gradients. In ice shelves a totally different situation exists. Here the basal shear stress is very small and a balance exists between normal stress gradient and hydrostatic pressure exerted by the sea water.

The derivation of (3.7.6) given above is based on physical intuition rather than mathematical rigor. A more careful analysis can be found in Budd (1970).

3.8   Ice shelves

So far we have only considered the flow in ice sheets resting on the bedrock. To conclude this chapter on ice-flow theory we now have a closer look at ice shelves. An ice shelf can be considered as the offshoot of an ice sheet and does not rest upon the bedrock (or only at some very limited places). A typical ice sheet – ice shelf system is sketched in Figure 3.7.

The equation for the normal stress deviator τ'_{xx} derived in the previous section enables us to calculate the steady-state profile of an unconfined free floating ice shelf. In this case (3.7.6) becomes

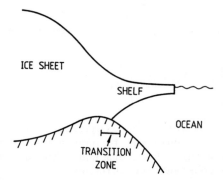

Figure 3.7. An ice sheet – ice shelf system. The ice starts to float when the surface elevation with respect to sea level is about 1/9 of the ice thickness. In the ice sheet the horizontal shearing stress is most important. In the ice shelf, on the other hand, the normal stress gradient balances the water pressure.

(3.8.1)
$$\frac{\partial}{\partial x} (2H\overline{\tau'_{xx}}) = \rho g H \frac{\partial h}{\partial x} \quad ,$$

where H is the total ice thickness and h the height of the ice surface above a reference level. Taking sea level as reference level and assuming hydrostatic equilibrium, the following relation holds:

(3.8.2)
$$h = (1 - \frac{\rho}{\rho_w}) H \quad .$$

The density of sea water is denoted by ρ_w. Substitution in (3.8.1) and integration with respect to x yields

(3.8.3)
$$2H\overline{\tau'_{xx}} = \frac{1}{2} \rho g (1 - \frac{\rho}{\rho_w}) H^2 + C \quad .$$

From the condition that the net normal force on the ice sheet should equal the hydrostatic pressure exterted by the sea water it follows that C = 0.

In a free floating ice shelf τ_{xx} plays a dominant role, which allows us to write the flow law as

(3.8.4)
$$\dot{\varepsilon}_{xx} = A \overline{\tau'_{xx}}^n = A^* \left[\frac{\overline{\tau'_{xx}}}{\tau_o} \right]^n \quad .$$

Combining this with (3.8.3) then gives an expression for the strain rate:

(3.8.5)
$$\dot{\varepsilon}_{xx} = A^* \left[\frac{\rho}{\rho_w} \frac{(\rho_w - \rho) g}{4 \tau_o} H \right]^n \quad .$$

We can make the perfect plasticity approximation by letting n go to infinity. In that case the only way to have a stable ice shelf is to set the expression within brackets equal to one. This yields an ice shelf with constant ice thickness H_p :

(3.8.6)
$$H_p = \frac{4 \tau_o \rho_w}{(\rho_w - \rho) g \rho}$$

Such an ice shelf can only exist when the mass balance at its surface is positive (stresses must continuously be built up). For a yield stress τ_o of 1 bar , (3.8.6) predicts a thickness of about 300 m.

It is obvious that an ice shelf of constant thickness is not realistic. The amount of ice entering the ice shelf at the grounding line should be important, and we can safely conclude that perfect plasticity theory is not useful for studying ice shelves. We therefore return to the flow law.

The strain rate can be expressed as

$$(3.8.7) \qquad \dot{\varepsilon}_{xx} = A \left[\frac{\rho g H (\rho_w - \rho)}{4 \rho_w} \right]^n = C H^n \ .$$

In order to find a steady-state profile we write the continuity equation (3.5.2) as

$$(3.8.8) \qquad \frac{\partial (Hu)}{\partial x} = H \dot{\varepsilon}_{xx} + u \frac{\partial H}{\partial x} = M \ .$$

Taking the accumulation or melting rate M constant, we see that the ice-mass flux can be written as

$$(3.8.9) \qquad Hu = Mx + H_o u_o \ ,$$

where x denotes the distance to the grounding line, and $H_o u_o$ the ice-mass discharge from the ice sheet at the grounding line. Multiplying (3.8.8) by H and inserting (3.8.7) and (3.8.9) yields

$$(3.8.10) \qquad (Mx + H_o u_o) \frac{\partial H}{\partial x} = MH - CH^{n+2} \ .$$

In the next section we will see how this equation can be solved and what the resulting equilibrium profiles look like.

Since most ice shelves are formed in embayments (the Ross Sea, for instance) we conclude this section by considering the equilibrium profile of an ice shelf confined by two mountain ranges at its sides. In such a situation lateral shear stresses exerted by the sides of the embayment on the shelf may become more important than the normal stress in the ice shelf.

If we denote the lateral shear stress by $\overline{\tau}_{xy}$, and neglect the normal stress in the ice shelf, stress equilibrium now becomes

$$(3.8.11) \qquad H \frac{\partial \overline{\tau}_{xy}}{\partial y} = \rho g H \frac{\partial h}{\partial x} \ .$$

With the use of (3.8.2) we obtain

$$(3.8.12) \qquad \frac{\partial \overline{\tau}_{xy}}{\partial y} = \rho g \left(1 - \frac{\rho}{\rho_w}\right) \frac{\partial H}{\partial x}$$

If the mountain ranges are roughly parallel to each other this equation can be integrated over the width W of the shelf to give

$$(3.8.13) \qquad \frac{\partial H}{\partial x} = \frac{\tau_s}{KW} \qquad \text{where } K = \rho g \left(1 - \frac{\rho}{\rho_w}\right) \quad,$$

and τ_s is the value of $\overline{\tau}_{xy}$ at the sides of the shelf. So for constant boundary conditions (W, τ_s) the slope of the ice shelf is now constant. For the Ross Ice Shelf typical values of τ_s are in the 0.4-1 bar range.

3.9 Equilibrium profiles of an unconfined ice shelf

In the preceding section an equation from which the equilibrium profile of a free floating ice shelf can be calculated, was derived. When the mass balance M is constant, the following relation holds:

$$(3.9.1) \qquad (Mx + H_0 U_0) \frac{\partial H}{\partial x} = MH - CH^{n+2} \quad .$$

In the simplest case (M = 0) this equation can be integrated immediately to yield the steady-state profile

$$(3.9.2) \qquad H = \left[\frac{(n+1)C}{H_0 U_0} x + H_0^{-(n+1)}\right]^{-1/(n+1)}$$

In this equation x denotes the distance to the grounding line and H_0 the ice thickness at x = 0.

   In the case of a non-zero mass balance the solution must fullfill the condition that $\partial H/\partial x$ is negative (for M = 0 this is always the case). The ice flow is in the direction of decreasing ice thickness, so whenever $\partial H/\partial x$ becomes positive the ice starts to flow towards the grounding line. Since this is, for a floating ice shelf, not a realistic situation, we require that $\partial H/\partial x < 0$. Consider first the case in which accumulation at the surface is larger than basal melting (M > 0). We rewrite (3.9.1) as

(3.9.3) $\dfrac{\partial H}{\partial x} = \dfrac{MH - CH^{n+2}}{Mx + H_0U_0}$.

Since $\partial H/\partial x$ must be negative, it follows that the ice thickness should be larger than some critical value H_{cr}, given by

(3.9.4) $H_{cr} = (M/C)^{+1/(n+1)}$.

When this condition is met, (3.9.3) can be integrated to give

(3.9.5) $H = \left[\dfrac{C}{M} - \dfrac{U_0^{n+1} (\frac{C}{M} H_0^{n+1} - 1)}{(Mx + H_0U_0)^{n+1}} \right]^{-1/(n+1)}$.

Most ice shelves are subject to basal melting. When the rate of melting exceeds accumulation at the surface ($M < 0$), we see from (3.9.3) that, since $\partial H/\partial x$ should be negative, the length of the ice shelf is restricted. The maximum length is given by

(3.9.6) $X_{cr} = - H_0U_0 /M$.

Figure 3.8 shows a comparison of ice-shelf profiles for different values of M. In all cases, in the vicinity of the grounding line the thickness of the shelf decreases rapidly.

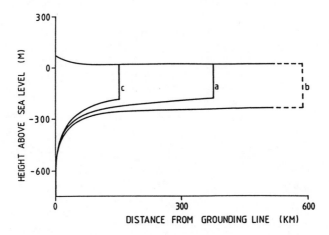

Figure 3.8. Steady-state profiles of an ice shelf. Profile a is calculated from (3.9.2), and profiles b (M = 0.15 m/yr) and c (M = -0.15 m/yr) from (3.9.5).

In case of zero mass balance (profile a), neither the thickness nor the length of the shelf is restricted. Therefore a critical thickness H_{cr} has to be used, below which all ice calves off instantaneously (otherwise the ice shelf would become infinitely long). In calculating profile a of Figure 3.8, H_{cr} was set to 200 m.

In case of a positive mass balance, the size of the ice shelf increases considerably. Since $\partial H/\partial x$ becomes smaller as the distance to the grounding line increases, the critical thickness H_{cr}, as given by (3.9.4), will be reached aymptotically. In reality this situation does not occur, of course, because at some point the mass balance will start to become negative.

We finally observe from Figure 3.8 that basal melting reduces the size of an ice shelf dramatically. In this case X_{cr} is 533 km, but to make a fair comparison between the profiles, the shelf was also cut off at the point where the thickness reached 200 m.

A more detailed discussion on the stress equilibrium in ice shelves can be found in Thomas (1973), Sanderson (1979) and Sanderson and Doake (1979).

4.   NUMERICAL ICE-SHEET MODELS

It is only in exceptional cases that the shape of an ice sheet
can be calculated by analytical methods. Some examples of
situations in which this is possible will be discussed in
chapter 6. In most cases numerical methods have to be used,
particularly when we are not only interested in steady states,
but also want to compute the transient behaviour of an ice sheet.
   In this chapter we discuss how numerical ice-sheet models,
based on the equations derived in previous sections, can be
constructed. As a start we consider a model based on a simple
vertically-integrated flow law, later we take up the more
difficult case in which normal stresses become important.

4.1 A model driven by shearing stresses only

The continuity equation, expressing conservation of ice volume
if ice density is taken constant, together with the flow law
(3.5.3) or (3.5.4) form a closed set of equations and can be
solved when proper boundary conditions are imposed. For instance,
combining (3.5.1) and (3.5.3) leads to an equation of the type

(4.1.1) $\dfrac{\partial H}{\partial t} = - \nabla \cdot (D \nabla h) + M$, where $H \geqslant 0$.

Here h is elevation of the ice surface above sea level, and the
factor D equals

(4.1.2) $D = B \; H^{m+1} \; (\nabla h \cdot \nabla h)^{\frac{1}{2}(m-1)}$.

So the evolution of the ice sheet is governed by a parabolic
differential equation. Equation (4.4.1) can be interpreted as a
diffusion equation for ice thickness, with a 'diffusivity' D
increasing strongly with ice thickness and surface slope (a
consequence of the non-linear character of the flow law).
   Parabolic differential equations are generally not very

67

difficult to solve. Stable schemes can easily be constructed (see e.g. Smith, 1978). For a linear diffusion equation of the type $\partial\chi/\partial t = D\nabla^2\chi$ the stability criterion for a forward-in-time/central-in-space difference scheme is given by

$$(4.1.3) \qquad \Delta t \leqslant \frac{(\Delta\ell)^2}{4\ D} \qquad .$$

Here Δt is the time step and $\Delta\ell$ the distance between grid points. From this condition it appears that the time step that can be used is generally small. A typical value of D in an ice sheet is $10^8 \text{m}^2/\text{yr}$, implying that with a grid spacing of 100 km the time step should be less than 25 yr.

For linear equations implicit schemes are more efficient than explicit schemes, because they are always stable and thus allow a larger time step. An implicit scheme requires the solution of a set of N linear equations in H_i , where N is the total number of grid points and H_i the ice thickness in the i-th grid point. If the matrix corresponding to the set of equations is inverted, the system is quickly solved at each time step. The problem with nonlinear equations now is that the elements of the matrix are not constant but change with time, implying that the matrix has to be inverted each time step. The net result is that for a nonlinear equation on a large grid implicit methods are hardly more efficient than explicit ones.

The main advantage of explicit schemes lies in the fact that coding is very easy and memory requirements can be kept to a minimum. Also, explicit schemes are much easier to adjust to changes in the boundary conditions and/or shape of the model domain. A scheme that performs well is the following one.
Let i denote the number of a grid point along the horizontal axis, and t the (discrete) time. $D_{i,t}$ is then first calculated from

$$(4.1.4) \qquad D_{i,t} = B_{i,t} \ H_{i,t}^{m+1} \ [\ (h_{i+1} - h_{i-1,t})/2\Delta\ell \ ]^{m-1}$$

Interpolated values of D are then used to construct the integration in time

$$(4.1.5) \qquad H_{i,t+1} = H_{i,t} + \frac{\Delta t}{2(\Delta\ell)^2} \left[(h_{i+1,t} - h_{i,t}) \right.$$

$$\left. (D_{i+1,t} - D_{i,t}) - (h_{i,t} - h_{i-1,t}) \ (D_{i,t} + D_{i-1,t}) \right]$$

$$+ M_{i,t} \ \Delta t \qquad .$$

It is of course a straightforward matter to write down the
corresponding difference equations for the two-dimensional
case (i.e., on a i,j-grid).
   Numerical experiments with (4.1.4)-(4.1.5) show that the
critical time step for stability is somewhat larger than the one
predicted by (4.1.3), provided that D is the maximum value of
$D_{i,j}$ over the grid. This observation implies the existence of an
initialization problem. In an ice sheet ice thickness and
surface slope are not independent. Smaller ice thickness is
generally associated with a larger surface slope. When H and h
are read from a map to a grid, truncation errors will be made.
Such errors may occasionally lead to very large values of D
and a numerical integration starting with the gridded ice-
thickness field may blow up immediately. This problem can be
solved by starting with a very small time step and increase it
later in the integration, or by simply setting an upper limit
to D until the model ice sheet has relaxed to the natural
balance between surface slope and ice thickness.

4.2 Boundary conditions

Numerical integration of the model described in the preceding
section requires proper boundary conditions. From (4.1.1) it
follows that if $M < 0$, the ice sheet cannot expand because $D = 0$
when $H = 0$. This is evidently an unrealistic situation and
results from a basic shortcoming of the model, namely, the
neglect of normal stress gradients. In spite of this a model
driven by shear stresses only appears to work. Due to truncation

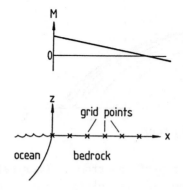

Figure 4.1. Geometry for a model experiment with a
flat continent bounded at one side.

errors in the numerical scheme, the divergence term $(-\nabla \cdot D\nabla h)$ at
the grid point i can be positive even if H_i is zero.
    To illustrate this point we consider a simple integration.
The model geometry for this experiment is shown in Figure 4.1.
It represents a flat continent bounded at one side by deep ocean
(here the boundary condition is H = 0). The mass balance is
prescribed to be a function of x only (an unrealistic situation,
in fact):

$$M = 0.5 - 0.5 \cdot 10^{-6} \, x \quad m/yr \quad ,$$

where x is in m. So one edge of the ice sheet will be at x = 0,
and in case of a steady state we expect the other edge close
to x = 1300 km. The integration starts with zero ice thickness
everywhere, and the spacing between the grid points is 100 km.
This rather large value makes it possible to see clearly the
consequences of spatial truncation errors.
    Calculated ice-sheet profiles are shown in Figure 4.2 at
intervals of 4000 yr. First there is no ice flow at all. When
the ice thickness is large enough, that is when stresses of
sufficient magnitude have been built up, the ice starts to flow
to the region where the mass balance is negative. Apparently,
the ice-sheet edge 'jumps' from grid point to grid point and is
not always so steep as it should be. Truncation errors can of
course be reduced by using a grid with higher resolution, but
still the ice-sheet edge will not move like a block. It appears

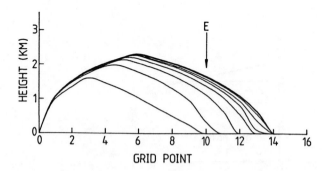

Figure 4.2. Ice-sheet profiles from a numerical model
integration, shown at 4000 yr intervals. The grid size
is 100 km. E indicates the equilibrium point (zero
mass balance).

that this does not seriously affect the evolution of the model
ice sheet. An integration with a 40 km grid, for instance,
gives virtually the same result as the integration displayed
in Figure 4.2. As long as the ice-sheet edge is located in a
region where the mass balance is negative, (4.1.1) produces a
well-defined ice-sheet edge. Or, to state it in a different way,
the size of the ice sheet is determined by its large-scale
mechanics and the mass balance, not by the small-scale mechanics
of the ice-sheet edge.

However, problems arise when the bedrock slopes steadily
downwards until a depth well below sea level (1000 m, say),
before the edge of the continental shelf is reached. When in
such a situation the ice-sheet edge reaches the coastline and
a floating ice condition is used (H is set to zero when the ice
is too thin to rest on the sea bed), the ice sheet cannot expand
further. This is a consequence of the nature of the diffusion
equation used to describe the ice-sheet evolution, and can be
understood as follows.

When $(\partial H/\partial t)\Delta t = \Delta H$ is calculated for the first grid point in
the shallow sea it will generally have a small value, even when
the ice thickness in the adjacent grid point on land is large.
For a time step of 10 yr, a typical value for ΔH is 5 m, which
obviously is not enough to overcome the floating ice condition
when the sea is deeper than a few meters. To handle such a
situation in a satisfactory way it is necessary to consider the
formation of an ice shelf. As is evident from section 3.8 this
involves a treatment of normal stresses as well. We take up this
point in the next section.

4.3 Modelling the ice sheet - ice shelf junction

Since the response time of an ice shelf is much smaller than the
characteristic time scale of ice-sheet growth or decay, we may
consider an ice shelf to be in equilibrium with the continental
ice sheet to which it is attached. So we can use the the model
described in the previous sections and add an ice shelf as soon
as the ice edge starts to float. However, in order to let the
ice sheet be able to expand, some assumption concerning the
profile of the snout has to be made. One approach is the
following.

Let i denote the last grid point where grounded ice is present,
and Δx the distance between two grid points. Assuming that the
snout has a parabolic profile with the top of the parabola
between the grid points i+1 and i+2, the ice thickness in i+1
equals

(4.3.1) $H_{i+1} = [\,(\Delta x/2)/(3\Delta x/2)\,]^{\frac{1}{2}} H_i = 0.577\ H_i$.

Given the bedrock topography and the parabolic profile of the ice-sheet edge, the position of the grounding line (the point where the ice starts to float) can be calculated. The ice thickness at the grounding line and the ice-mass discharge from the sheet are then known, and (3.9.2) directly gives the profile of the ice shelf.

When the ice shelf experiences no friction at its sides, the backward pressure exerted by the ice shelf on the grounded ice will be small. However, when the ice shelf has run aground or is in another way restricted in its movement, we expect a considerable effect on the grounded ice sheet. To study this in more detail it is necessary to consider normal stresses in the vicinity of the grounding line. These normal stresses tend to reduce the effective basal shear stress near the grounding line, in the so-called transition zone.

We start by parameterizing the basal shear stress in the transition zone as

$$(4.3.2) \qquad \tau_b(x) = \tau_b(-L_1) - (x+L_1)\,\tau_b^{*} \quad .$$

Here L_1 denotes the distance to the grounding line at which normal stress gradients are assumed to become unimportant. For convenience we have set $x = 0$ at the grounding line; see Figure 4.4 for the geometry.

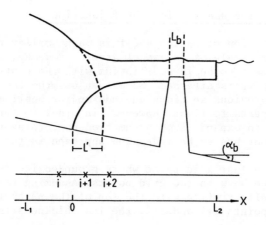

Figure 4.4. Geometry used in formulating the model of the ice sheet - ice shelf junction.

In section 3.7 the following balance of forces acting on a column of ice was discussed:

$$(4.3.3) \qquad \frac{\partial}{\partial x} (2H \overline{\tau'_{xx}}) = \rho g H \frac{\partial h}{\partial x} + \tau_b .$$

In order to find an expression for τ_b^{*}, we integrate (4.3.3) from $x = -L_1$ to the end of the ice shelf at $x = L_2$, thereby ensuring an overall stress equilibrium. Using (3.8.3) at the ice-shelf edge, the parabolic profile $H = \sigma(L'-x)^{\frac{1}{2}}$ for the snout of the ice sheet, and (3.9.2) for the ice-shelf profile then yields the following integrated stress equilibrium:

$$(4.3.4) \qquad \frac{1}{2} \rho g H_e^2 (1-\frac{\rho}{\rho_w}) = \frac{1}{2} \rho g \sigma^2 L_1 \qquad (B)$$

$$- \frac{1}{3} \rho g \sigma \alpha_b \{L'^{3/2} - (L'+L_1)^{3/2}\} \qquad (C)$$

$$+ \frac{1}{2} \rho g (1-\frac{\rho}{\rho_w}) \left[\{\frac{C}{U_o H_o} L_2 + H_o^{-n-1}\}^{-2/(n+1)} - H_o^2 \right] \qquad (D)$$

$$+ L_1 \tau_b(-L_1) - \frac{1}{2} \tau_b^{*} L_1^2 \qquad (E)$$

$$+ \overline{\tau}_b L_b \qquad (F)$$

In this expression H_o is the ice thickness at the grounding line, H_e the (prescribed) ice thickness at the edge of the shelf, U_o the ice velocity at the grounding line and α_b the slope of the bedrock. The left-hand side of (4.3.4) is the force exerted by the sea water on the edge of the ice shelf. The terms on the right-hand side can be given the following interpretation:
(B): pressure gradient force arising from the slope of the ice-sheet edge;
(C): correction for the slope of the bedrock under the snout of the ice sheet;
(D): pressure gradient force due to the slope of the ice-shelf surface;
(E): integrated reduced basal shear stress in the transition zone;
(F): backward force exerted by the ice shelf when it has run aground.
In the situation shown in Figure 4.4 the most important terms are (E) and (F), having an order of magnitude of 10^{10} kg/s^2. So when the ice shelf runs aground the backward pressure will

essentially be compensated by a reduction of the basal shear
stress in the transition zone.

By writing down (4.3.2), we have introduced a new quantity L_1,
the length of the transition zone. The value of L_1 is not well
known, and we may even expect it to be a function of the ice-
shelf geometry itself. In the examples discussed in the next
section the value of L_1 was set to 100 km.

We are now able to compose a 'complete' model of an ice sheet -
ice shelf system. For a given state of the ice sheet, τ_b^* can be
calculated from (4.3.4). From (4.3.2) the distribution of the
effective basal shear stress is then obtained directly, and the
evolution of the ice sheet can be computed in the usual way (see
sections 4.1 and 4.2). This gives new values of H_O and U_O, and
the procedure can be repeated. In the next section we discuss
a few examples of numerical integrations with this scheme.

4.4   An ice sheet on a sloping bedrock

The scheme described in the previous section makes it possible
that a model ice sheet expands further even when the bed is
considerably below sea level. Also, it is possible now to
demonstrate what happens with the main ice sheet when the ice
shelf runs aground.

Figure 4.5 shows a comparison of two steady-state profiles,
one for a bed without and one for a bed with a seamount. In
both cases, the slope of the bedrock is 0.001 and the mass
balance 0.1 m ice depth/yr. The profiles were obtained by
running the model until stationary conditions were established.

In the upper part of the figure the case with a free floating
ice shelf is shown. The ice sheet is able to extend far in the
ocean, with an ice thickness of about 2 km at the grounding
line. This is more than generally observed, and the reason
probably is that in the present example sliding was not
considered (the importance of sliding with regard to grounding-
line migration will be discussed in chapter 11). In the example
shown here, the length of the shelf is determined by the critical
thickness H_{cr}, set to 250 m. Taking a value of 200 m would lead
to an ice shelf four times as long.

The lower part of Figure 4.5 shows the case in which the
ice shelf runs aground on a bedrock irregularity. In such a
situation the choice of H_{cr} is very critical, of course. It
determines whether the ice shelf is long enough to reach the
seamount. When it does, the last term in (4.3.4) becomes
important and the ice shelf exerts backward pressure to the
ice sheet. This enables the ice sheet to grow past the bedrock
irregularity.

In these examples the length of the transition zone was set

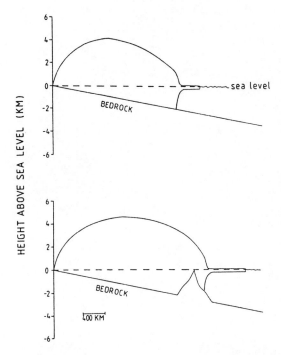

Figure 4.5. Steady-state ice-sheet profiles illustratin'
the effect of a bedrock irregularity on which an ice
shelf can run aground.

to 100 km. Since we do not want the behaviour of the ice sheet
to depend to much on the grid actually used, the transition zone
should included several grid points. So the grid-point spacing
should not be larger than 40 km, say.

5. THE HEAT BUDGET OF ICE SHEETS

In this chapter we will discuss the distribution of temperature in an ice sheet. This topic deserves attention because the rate of deformation depends strongly on temperature. Moreover, the presence of basal water can lead to sliding. Melt water then acts as lubricator and the horizontal ice-mass discharge may increase dramatically. In order to calculate the amount of melt water that is produced, the heat budget of the ice sheet has to be known.

5.1 The thermodynamic equation

To find a suitable expression for how ice temperature changes with time, we start with the first law of thermodynamics. It reads (per unit of mass):

$$(5.1.1) \qquad \frac{dE}{dt} = \frac{dQ}{dt} + (p/\rho^2) \frac{d\rho}{dt} \quad .$$

Here E is the internal energy, p hydrostatic pressure and ρ density. The first term on the right-hand side represents the diabatic heating.

Using some elementary thermodynamic relations we have

$$(5.1.2) \qquad \frac{dE}{dt} = (\partial E/\partial T)_p \frac{dT}{dt} + (\partial E/\partial p)_T \frac{dp}{dt} =$$

$$= (c_p - \alpha p/\rho) \frac{dT}{dt} + (\beta p/\rho - \alpha T/\rho) \frac{dp}{dt} \quad ,$$

and

$$(5.1.3) \qquad \frac{d\rho}{dt} = (\partial \rho/\partial T)_p \frac{dT}{dt} + (\partial \rho/\partial p)_T \frac{dp}{dt} =$$

$$= -\rho\alpha \frac{dT}{dt} - \rho\beta \frac{dp}{dt} \quad .$$

77

Here α is the coefficient of expansion and β the isothermal compressibility:

$$\alpha = -\frac{1}{\rho} (\partial\rho/\partial T)_p \quad , \quad \beta = -\frac{1}{\rho} (\partial\rho/\partial p)_T \quad .$$

The heat capacity at constant pressure is denoted by c_p.

The heat supplied to a unit of mass is due to divergence of the conductive heat flux $(\nabla\cdot\vec{F})$, viscous dissipation of mechanical energy (Φ) and external sources or sinks (J). If k is the thermal conductivity of ice (which we assume to be constant), the conductive heat flux equals $-k\nabla T$. The rate of heating can thus be written as

(5.1.4) $$\frac{dQ}{dt} = \frac{1}{\rho} (-\nabla\cdot\vec{F} + J + \Phi) = \frac{1}{\rho} (k\nabla^2 T + J + \Phi) \quad .$$

Substitution of (5.1.2) - (5.1.4) into (5.1.1) leads to the required equation for the rate of change of temperature:

(5.1.4) $$c_p \frac{dT}{dt} - \frac{\alpha T}{\rho} \frac{dp}{dt} = \frac{1}{\rho} (k\nabla^2 T + J + \Phi).$$

In an ice sheet the second term on the left-hand side can be neglected, except in the snow - ice transition layer near the surface.

Since the only heating associated with external processes results from freezing of water or melting of ice, we set J equal to $W/\rho c_p$, where W is the latent heat of fusion. Combining this with an Eulerian description in which the total derivative dT/dt is split into a local derivative and an advection term, the temperature equation becomes

(5.1.6) $$\frac{\partial T}{\partial t} = \frac{k}{\rho c_p} \nabla^2 T - \vec{v}\cdot\nabla T + \frac{W}{\rho c_p} + \Phi \quad .$$

Finally, the internal heating due to deformation can be expressed as (e.g. Paterson, 1981, p. 199):

(5.1.7) $$\Phi = \sum_{ij} \dot{\varepsilon}_{ij} \tau_{ij} \quad .$$

In general (5.1.6) cannot be solved by simple methods. However, with a number of simplifying assumptions the basic features of the temperature field in an ice sheet can be revealed.

5.2   Some basic solutions

The thermodynamic equation derived in the previous section can
only be solved analytically by making extensive simplifications.
A method applicable to the central part of an ice sheet was
given by Robin (1955). He made the following assumptions:
- the basal temperature is below the melting point of ice;
- the horizontal temperature gradient is small compared to
  the vertical temperature gradient, so $\nabla^2 T = \partial^2 T/\partial z^2$;
- the horizontal advection of heat is negligible ;
- dissipative heating is mainly in the lower layers of the
  ice sheet and can be taken into account by increasing the
  geothermal heat flux.

In this case the thermodynamic equation takes the form

$$(5.2.1) \qquad \frac{\partial T}{\partial t} = \frac{k}{\rho c_p}\frac{\partial^2 T}{\partial z^2} - w\frac{\partial T}{\partial z} \ .$$

In order to solve this equation we further consider steady
states, which makes it possible to relate the vertical
velocity w to the ice-accumulation rate M. Without freezing or
melting at the base the simplest possible velocity profile then
is

$$(5.2.2) \qquad w(z) = -Mz/H \ .$$

Steady-state temperature profiles can thus be found by solving

$$(5.2.3) \qquad \frac{k}{\rho c_p}\frac{\partial^2 T}{\partial z^2} + \frac{Mz}{H}\frac{\partial T}{\partial z} = 0 \ .$$

As boundary conditions we have: (i) the temperature at the
surface of the ice sheet is constant, so $T(H) = T_s$; (ii) the
temperature gradient at the base matches the prescribed (and
eventually corrected) geothermal heat flux ψ, so $(\partial T/\partial z)_b = -\psi/k$.
Integrating (5.2.3) twice yields

$$(5.2.4) \qquad T-T_b = (\partial T/\partial z)_b \int_0^z \exp(-z^2/q^2)dz \ ,$$

$$\text{where } q^2 = \frac{2kH}{\rho c_p M} \ .$$

When the accumulation rate M is positive, this can also be
written as

(5.2.5) $T-T_s = \frac{1}{2}\sqrt{\pi}\,(\partial T/\partial z)_b\,\{erf(z/q) - erf(H/q)\}q$.

The error function is defined as

$$erf\ z = \frac{2}{\sqrt{\pi}} \int_0^z exp(-z'^2)dz'\ .$$

Tables of this function can be found in most mathematical handbooks (e.g. Abramowitz and Stegun, 1965).

   Figure 5.1 shows the solution (5.2.5) for an ice sheet with a thickness of 2500 m. Profiles are shown for various values of the ice-accumulation rate M. Numerical values of the constants used in the calculation are: $k/\rho c_p = 1.15 \times 10^{-6}$ m^2/s and $(\partial T/\partial z)_b = -0.0167$ K/m (corresponding to a geothermal heat flux of 4.2×10^{-2} W/m^2 and an additional equally large temperature gradient reflecting the effect of frictional heating). From the figure it is immediately clear that the accumulation rate has a very strong influence on the steady-state temperature profile. If M=0 the temperature profile is a straight line with a slope equal to the geothermal gradient $(\partial T/\partial z)_b$. For values of M larger than 0.3 m/yr the upper half of the ice sheet becomes isothermal as a result of the downward ice flow that carries cold ice from the surface to larger depth.

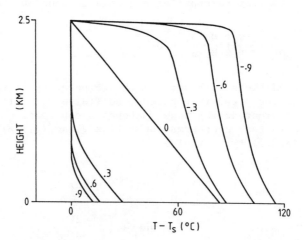

Figure 5.1. Temperature profiles (steady state) for an ice sheet which is 2500 m thick. Labels give the accumulation rate at the surface in m ice depth / yr.

The profiles for negative values of M (ablation) were
computed numerically by integrating the thermodynamic equation
until a steady state was reached (this takes about 20 000 years
of simulated time). Apparently a negative mass balance leads to
a much warmer ice sheet. This is due to the upward ice flow that
brings ice from the lower and warmer part of the ice sheet to
the surface. We cannot expect (5.2.3) to be valid in this
situation because the ice would be above the melting point
everywhere for realistic surface temperatures.

How horizontal advection of heat affects the temperature
profile becomes clear when we consider stream lines in an ice
sheet. Consider for instance the column AB in Figure 5.2.
Following the ice-particle paths upstream we see that ice at a
deeper location originates from a higher surface where
temperature is lower. As shown in the right of Figure 5.2, this
implies that ice temperature decreases with depth until the
geothermal heat flux becomes dominant. Such profiles, with a
temperature inversion in the upper layer of the ice sheet, can
only be found if in the thermodynamic equation the horizontal
advection term is taken into account. This will be done in a
later section.

Assuming that the geothermal heat flux is fairly constant
over very long periods of time, variations of the temperature
distribtution in an ice sheet will be initiated at the surface
by either a change in surface temperature or a change in the
mass balance. So it is of some importance to see how a
temperature perturbation at the surface travels downward.
To investigate this point we consider a variation of the
surface temperature which is sinusoidal in time, i.e.

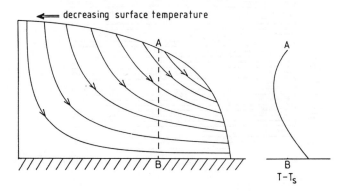

Figure 5.2. Reversal of temperature gradient in the
upper layers when advection becomes important.

(5.2.6) $T(H,t) = T_s \sin(\omega t)$.

If, for a moment, we only consider diffusion (w=0), the time-dependent solution for the temperature profile becomes

(5.2.7) $T(z,t) = T_s \exp(-z') \sin(\omega t - z')$,

$$\text{where } z' = (H-z) \left[\frac{w\rho c_p}{2k} \right]^{1/2}$$

So the amplitude of the temperature wave decreases as exp(-z') implying that the higher the frequency, the more rapidly attenuation with depth takes place. Temperature perturbations with a long time scale thus propagate further downwards than those with a short time scale.

More precisely, the depth at which the maximum temperature change is 5% of T_s is given by

(5.2.8) $H-z = \left[\dfrac{2k}{w\rho c_p} \right]^{1/2} \ln(20)$.

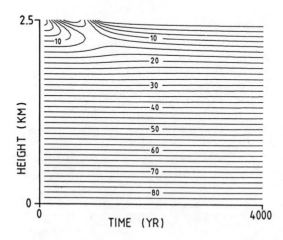

Figure 5.3. Downward penetration of a perturbation in surface temperature by conduction only. Temperature is given with respect to undisturbed surface temperature (K).

This maximum occurs at a time $t = \ln(20)/\omega$ after the surface temperature has reached its maximum value. If T_s has a period of one year (the annual cycle), the 5 % depth is 10 m and the maximum temperature at this depth occurs $5\frac{1}{2}$ month later. It should be noted, however, that in reality downward diffusion of a temperature perturbation takes even more time, because the thermal conductivity of snow and firn is normally smaller than that of ice. In the analysis given above, it has been assumed that the ice sheet consists of solid ice only.

As an illustration to the solution (5.2.7), Figure 5.3 shows how a temperature wave (with a duration of half the period) travels downward into the ice sheet. The period was set to 500 yr and the amplitude to 20 K. Apart from the unrealistically large temperature difference between surface and base (because there is no downward advection) the penetration depth apparently is very small.

To see clearly the essential role played by downward advection of cold ice, another (numerical) calculation was carried out with a downward vertical velocity at the surface of 0.3 m/yr, decreasing linearly with depth. The temperature perturbation imposed was the same. As shown in Figure 5.4 the difference is striking. The temperature wave travels much further: it is advected downward before it has been 'diffused away'. On the basis of this result we may state that the response of an ice sheet to a change in surface conditions very much depends on the ice-accumulation rate, and will therefore vary considerably from place to place.

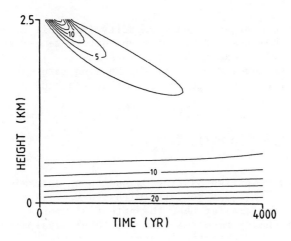

Figure 5.4. As in Figure 5.3, now taking into account downward advection.

5.3   The total heat budget of an ice sheet

Although the dynamical behaviour of an ice sheet depends on the structure of the temperature field, it is instructuve to have a look at the total heat budget.

As a basic simplification we assume that the vertical temperature gradient in the upper ice layers is very small, implying that diffusive heat transfer to the atmosphere is negligibly small. This is a plausible assumption for polar ice sheets which are hardly subject to melting at the surface. In a steady state, a balance thus exists between dissipative heating, geothermal heat entering at the base, and advection of heat (by means of snow accumulation at the surface and calving at the ice-sheet edge). From this balance it is possible to calculate a characteristic temperature of the ice leaving the ice sheet at the edge.

We consider one half of a one-dimensional ice sheet. Denoting its half volume by V and its mean thickness by H, the geothermal heat input is (the length of the sheet is V/H)

$$(5.3.1) \qquad Q_g = VG/H \quad .$$

Here G is the geothermal heat flux.

Advection is made up of two contributions. The heat input associated with accumulation at the surface equals

$$(5.3.2) \qquad Q_a = \rho c M V (T_s - \bar{T})/H \quad .$$

T is mean ice temperature, c heat capacity (per unit of mass), M the mean mass balance and T_s the mean surface temperature. Expression (5.3.2) is exact when the mass balance is independent of surface temperature. Similarly, the heating rate due to calving at the ice-sheet edge is (note that a steady state is considered)

$$(5.3.3) \qquad Q_e = - \rho c M V (T_e - \bar{T})/H \quad .$$

Finally the dissipative heating Q_d has to be determined. Since the kinetic energy associated with ice flow is several orders of magnitude smaller than the potential energy, conservation of energy requires that the total dissipation equals the release of potential energy by downward motion. This argument can be used to calculate Q_d. Per unit of time an amount of MV/H of ice mass flows through the ice sheet. The associated loss of potential energy (ρgH per unit width) is

$$(5.3.4) \qquad Q_d = gMV \quad .$$

The total heat budget of the ice sheet now reads

$$(5.3.5) \qquad Q_g + Q_a + Q_e + Q_d =$$

$$\frac{V}{H} \{G + \rho cM(T_s - T_e)\} + \rho gMV = 0 \quad .$$

Given the mass balance, mean surface temperature, ice thickness and volume (or length), the ice temperature at the edge of the ice sheet can be calculated from this equation.

Equation (5.3.5) obtains a wider applicability when a relation between mass balance and ice volume can be added. Such a relation can be derived by inserting typical quantities (a length scale L, mean mass balance M, typical ice thickness H) in the vertical-mean ice flow model (4.1.1)-(4.1.2). This procedure is discussed in more detail in section 8.3). The resulting balance between snow accumulation and ice-mass discharge reads

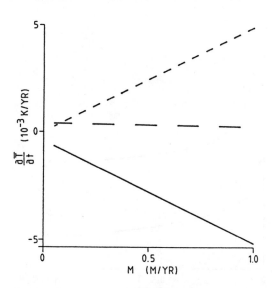

Figure 5.5. Components of the heat budget for an ice sheet of fixed length (1000 km), as a function of the mass balance. The lines show dissipation (-----), geothermal heat (– – – –) and advection term (———).

(5.3.6) $C H^{2m+1} L^{-m} = M L$.

Here C is a generalized flow parameter. This equation directly gives the mean ice thickness (and volume) when the length of a drainage system is prescribed and constant. Combining (5.3.6) and (5.3.5) now allows a calculation of the heat budget of an ice sheet in dependence of the mass balance.

Figure 5.5 shows the result of such a calculation. The various terms of the heat budget are given as a function of M, for an ice sheet with a typical half-width of 1000 km. Values of the flow parameters used are: m=3, C=1 $(m^2 yr)^{-1}$. Surface temperature was assumed to decrease linearly with elevation:

$$T_s = 250 - 0.009 H \quad K \quad .$$

A low sea-level temperature of 250 K was used to be sure that T_e would not reach the melting point.

All terms in the heat budget equation were divided by the heat capacity of the ice sheet $(\rho c V)$, so the vertical scale has unit K/yr. From the figure we see that geothermal heat is only important when the mean accumulation rate is low, less than 0.3 m ice depth per year, say. For large accumulation rates a close balance between dissipation and advection exists.

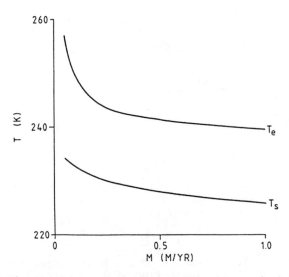

Figure 5.6. Mean surface temperature (T_s) and ice temperature at the ice-sheet edge (T_e), corresponding to Figure 5.5. From Van der Veen and Oerlemans (1984).

Corresponding values of T_s and T_e are shown in Figure 5.6. Note that the T_s curve directly reflects changes in the ice thickness. For M < 0.2 m/yr, the characteristic ice temperature at the edge is extremely sensitive to M: it rapidly drops when accumulation increases. This can be understood by realizing that the mean residence time of ice in the ice sheet increases when M decreases, implying that an ice column can take up geothermal heat during a longer period of time.

Another possibility of the approach presented in this section is to calculate the production of basal melt water. For high sea-level temperatures and sufficiently low accumulation rates, T_e will reach the melting point and the heat budget can only be zero when heat is lost by melting. The rate of melting follows from the imbalance in the heat budget. A further discussion on this is given in the chapter on interaction of ice flow and the heat budget.

5.4   An approximate method to calculate the temperature field

In the foregoing sections we have discussed some special solutions of the thermodynamic equation, which have, however, restricted applications. For general purposes an explicit solution can only be obtained by numerical methods.

The thermodynamic equation is of a mixed diffusive-advective type, and can be solved by standard methods (e.g. Mesinger and Arakawa, 1976). To do this on a three-dimensional grid has some disadvantages. First of all the required computational times are very large, too large for use in paleoclimatic studies. Another problem concerns the variable dimensions of the domain (i.e. the ice sheet). In particular changing ice thickness is difficult to deal with, because the upper boundary condition varies strongly with height. So a lot of interpolation is required, or the ice thickness should be restricted in such a way that the surface is always at a grid point. Even then horizontal advection of heat still gives problems.

One way to circumvent the difficulties associated with discretization in the vertical is to scale the vertical coordinate by the ice thickness. This approach was investigated by Jenssen (1977). He transformed the governing equations, yielding a scheme in which the advective terms can be dealt with much easier. Nevertheless, the method appeared to be moderately succesful because of large computational times and numerical instabilities appearing the ice-sheet edge.

A more efficient method can be designed by using a spectral approach, i.e. by expanding the vertical temperature profile in terms of functions of z, the height above the bedrock. The motivation for this is that temperature profiles in ice sheets are rather smooth, and can thus be described by a few components.

So we write

(5.4.1) $T(x,y,z,t) = \Theta_0(x,y,t) + z\,\Theta_1(x,y,t) +$

$+ z^2\,\Theta_2(x,y,t) + \dots$

Three terms are retained, so three equations are needed for the spectral coefficients Θ_0, Θ_1, and Θ_2. These can be obtained from the lower and upper boundary condition, and from the vertically-integrated heat equation.

As lower boundary condition we again prescribe the geothermal heat flux ψ, yielding

(5.4.2) $\Theta_1 = -\,\psi/(\rho\ell c)$,

where ℓ is the thermal diffusivity. At the surface (z=H) the ice temperature is set equal to the annual air temperature T_s. This gives

(5.4.3) $\Theta_2 = (T_s - \Theta_0 + GH/\ell)/H^2$.

For convenience $\psi/\rho c$ has been replaced by G, which is considered to be constant.

To obtain the third equation we use (5.1.6). Disregarding the effect of melting or freezing, we find by integrating from the bottom (z=0) to the surface (z=H):

(5.4.4) $\displaystyle\int_0^H \frac{\partial T}{\partial t}\,dz = (\partial T/\partial z)_H + G - \int_0^H w\,\frac{\partial T}{\partial z}\,dz\,-$

$\displaystyle -\int_0^H \left\{u\frac{\partial T}{\partial x} + v\frac{\partial T}{\partial y}\right\}\,dz + \int_0^H \Phi\,dz$.

Here u and v are the components of the velocity vector $\vec{v}$ parallel to the bedrock. Note that diffusion in the horizontal direction has been neglected.

An evaluation of the advective contribution requires specification of the velocity profile. A reasonable assumption for the horizontal velocity is

(5.4.5) $\vec{v} = \frac{3}{2}\,H^{-3/2}\,\sqrt{z}\,\,\vec{\bar{V}}$,

where $\vec{\bar{V}}$ is the vertical mean horizontal ice velocity as obtained

from a flow law of the type discussed in section 3.5. So the velocity profile is prescribed to be parabolic, but this is not crucial for the approach presented here.
Continuity of mass requires

$$\frac{\partial w}{\partial z} = - \nabla \cdot \vec{v} \simeq \frac{3}{2} H^{-3/2} \sqrt{z} \; \nabla \cdot \vec{v} \; .$$

In this equation the right-hand side is the divergence of the velocity field parallel to the bedrock. Obviously, the vertical velocity profile compatible with (5.4.5) should read

(5.4.6) $w = W_H \, (z/H)^{3/2} \; .$

W_H denotes the vertical velocity at the surface of the ice sheet.
    Inserting (5.4.5) and (5.4.6) in the heat equation (5.4.4) yields, after integrating over the vertical and some manipulation to eliminate Θ_1 and Θ_2:

(5.4.7) $\frac{2}{3} H \frac{\partial \Theta_0}{\partial t} = \Theta_0 \left(\frac{4}{7} W_H - 2\ell/H - \frac{2}{3} \frac{\partial H}{\partial t}\right) +$

$\qquad\qquad + \frac{\partial H}{\partial t} \left(\frac{2}{3} T_s + \frac{1}{3} GH/\ell\right) - \frac{1}{3} H \frac{\partial T_s}{\partial t} + 2G +$

$\qquad\qquad + T_s (2\ell/H - \frac{4}{7} W_H) - \frac{6}{35} W_H \, GH/\ell -$

$\qquad\qquad - UH \frac{\partial \Theta_0}{\partial x} - \frac{3}{7} UH^3 \frac{\partial}{\partial x} \left\{\frac{1}{H^2}(T_s - \Theta_0 + GH/\ell)\right\} +$

$\qquad\qquad + \int\limits_0^H \Phi dz \; .$

This equation is valid for the two-dimensional case (height and length). For the three-dimensional case a line of the type of the last but one in (5.4.7) should be added to deal with horizontal advection in the y-direction. In the following we will treat the two-dimensional case for simplicity.
    One of the reasons that (5.4.7) appears as a rather complicated equation is the temporal variability of the domain (H and T_s are functions of time, in general). Nevertheless, using (5.4.7) is still much more efficient than employing a vertical grid.
    The boundary conditions require the vertical velocity at the surface to be known. This vertical velocity is related to the change in ice thickness according to

$$(5.4.8) \qquad W_H = \frac{dH}{dt} - M = \frac{\partial H}{\partial t} + U_H \frac{\partial H}{\partial x} - M \quad .$$

M is the rate of accumulation, U_H the horizontal ice velocity at the surface (which follows from the vertical mean velocity and the assumption concerning the velocity profile). Since melting or freezing at the base is generally small, $w(z=0)=0$ is a sufficiently accurate lower boundary condition. Note that in the formulation used here downward velocity is negative.

Next we turn to the last term in (5.4.7), the frictional heating. Again, since in an ice sheet the kinetic energy is very small, the release of potential energy by downward motion should equal the dissipative heating. At this point it becomes important to distinguish between vertical velocity relative to the bedrock topography (which has been used so far), and vertical velocity in an absolute frame of reference, i.e. attached to the geoid. A completely rigid slab of ice sliding down a slope illustrates this point: although there is no deformation and the relative vertical velocity is zero, a lot of potential energy is dissipated.

It is also important to realize that dissipation does not equal the time rate of change of potential energy, because changes in surface elevation due to accumulation or melting are not associated with dissipation (here the energy budget of the atmosphere comes into play). So we obtain

$$(5.4.9) \qquad \int_0^H \Phi \, dz = - \frac{1}{\rho c} \int_0^H \rho g w' \, dz = - \frac{gH}{c} \{ \frac{2}{5} W_H + U \frac{\partial b}{\partial x} \} \quad ,$$

where we have used (5.4.7) and

$$(5.4.10) \qquad w' = w + u \frac{\partial b}{\partial x} \quad .$$

In these equations the bedrock topography is denoted by b.

In order to see the importance of the various terms in the heat budget, we now combine the treatment of thermodynamics given above with output from a vertically-integrated ice flow model. At this point we do not yet take into account the temperature – ice flow feedback.

First, the evolution of a northern hemisphere ice sheet was calculated with (4.1.1) and (4.1.2), with an ice-accumulation rate according to

$$(5.4.11) \qquad M = \min \{0.4, \ 0.0012 \ (h-h_s) + 0.7 \ x\} \quad m/yr \quad .$$

Here h_s is the height of the snow line and h the surface elevation, which equals the ice thickness since bedrock sinking is ignored. Starting with zero ice volume a steady state is reached in about 50 000 yr. The corresponding ice thickness and velocity are shown in Figure 5.7. During the numerical integration $H(x,t)$ and $U(x,t)$ were stored, and could thus be used later as input for the thermodynamic calculations.

Before (5.4.7) can be integrated in time, the temperature at the surface of the ice sheet has to be specified. It seems natural to prescribe the temperature at the snow line (-12 °C in the present example) and calculate the surface temperature T_s by using a constant lapse rate (-9 K/km). All time-dependent boundary conditions being known, the temperature field can now be calculated by numerical integration of (5.4.7), using (5.4.8) and (5.4.9). To gain a maximum of insight the best procedure is to include the various terms step by step. Figure 5.8 summarizes the results in terms of basal temperature in a running-time diagram.

Panel A shows a calculation in which only downward advection and diffusion are taken into account. Basal melting is ignored completely. The lowest values of Θ_0 are found just south of the northern edge of the ice sheet, because at this location surface temperature is rather low while the ice thickness is small. The evolution to a steady state is quite regularly, it is reached after about 60 000 yr.

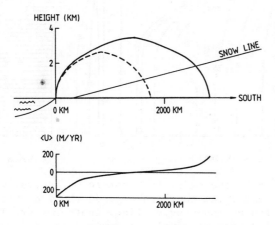

Figure 5.7. Geometry of the northern hemisphere ice-sheet model used in the thermodynamic calculation. The solid line gives the steady-state profile, the dashed line the profile after 20 000 yr. The lower curve shows mean ice velocity in the steady state.

Panel B gives the solution for the case in which frictional heating is included. The effect is most pronounced halfway between the centre and the northern edge of the ice sheet. As a consequence, the north - south gradient in basal temperature decreases. If horizontal advection of heat is included, the picture changes dramatically. As shown in panel C, the advection of cold ice towards the edges creates zones where the basal

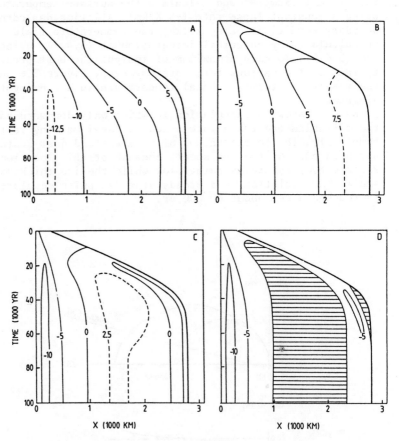

Figure 5.8. The basal temperature Θ_0 according to some experiments. The southern edge of the ice sheet is indicated by the heavier line. Isotherms are labeled in $^\circ$C. The different panels correspond to: (A) only conduction and downward advection taken into account, (B) frictional heating is added, (C) horizontal advection of heat is included, (D) basal melting is dealt with. In the calculations the geothermal heat flux was set to 0.042 W/m^2.

temperature is comparatively low. The lowest temperatures are found close to the northern edge.

Finally, if basal melting is taken into account panel D emerges. The rate of melting has been calculated by keeping Θ_0 at the melting point (corrected for pressure) and use the surplus heat for melting. The regions were melting occurs are shaded. Melting rates are small: typically a few mm per year. It can be seen that the corresponding lower basal temperature in the central part of the ice sheet causes lower basal temperatures near the southern edge through the horizontal advection process.

A comment on the numerical scheme used to integrate (5.4.7) is in order. The explicit scheme that performs best is the Lax-Wendroff scheme (e.g. Mesinger and Arakawa, 1976), which uses forward time-differencing and includes a (diffusive-type) correction for the advective terms. For the pure advection equation $\partial Y/\partial t + U\partial Y/\partial x = 0$ this scheme reads

$$(5.4.12) \qquad Y_i(t+\delta t) = Y_i(t) - \frac{U\delta t}{2\delta x}\{Y_{i+1}(t) - Y_{i-1}(t)\} +$$

$$+ \frac{1}{2}(U\delta t/\delta x)^2 \{Y_{i+1}(t) - 2Y_i(t) + Y_{i-1}(t)\} \quad .$$

Here the second line represents the artificial diffusion, which makes the scheme stable. Nevertheless, when the horizontal temperature gradient near the ice-sheet edge becomes very large, numerical instability may occur. This is for example the case when the edge of the ice sheet advances rapidly. In such situations reduction of the time step may help. Also, setting an upper limit to the temperature gradient near the edge may stabilize a 'wild snout'.

While maintaining the basic idea of representing the vertical temperature profile by a second-order polynomal, refinements to the scheme described here can be made. For instance, the dissipative heating can be split into a contribution to the mean temperature of the ice column and a contribution to the basal temperature gradient. Another possibility is to include a calculation of the bedrock temperature. Then the lower boundary condition for the ice temperature profile should be formulated in terms of basal temperature rather than basal temperature gradient.

This brings us at the end of a somewhat lengthy chapter. Not much work was quoted, we therefore give a few additional references here. Weertman (1968) proposed a simple model to calculate a local temperature profile taking into account the effect of horizontal advection. A more dynamic approach was given by

Budd et al. (1971), the so-called moving column model, in which a column of ice is followed during its travel along a flow line. Time-dependent integrations to study the sensitivity of a temperature profile in a single column were carried out by many workers, e.g. Budd et al. (1976). To our knowledge, the only calculation on a fully three-dimensional grid was the one by Jenssen (1977), already discussed in the beginning of this section.

6. INTERACTION OF ICE FLOW AND HEAT BUDGET

In this chapter we discuss how the temperature field affects the ice-mass discharge. After consideration of the local coupling between strain heating and deformation rate, we turn to a discussion whether ice sheets may acquire different steady-state profiles for identical environmental conditions. To this end we use both simple integrated models (considering the total heat budget again), and more explicit calculations with a numerical model.

6.1 Temperature dependence of the flow parameter

In section 3.1 we discussed the general flow law

$$(6.1.1) \qquad \dot{\epsilon}_{ij} = A \, \tau^{n-1} \, \tau'_{ij} \, .$$

When the temperature in an ice sheet is not calculated explicitly, the parameters A and n have to be taken constant. Their values are then chosen in such a way that realistic results are obtained. For example, using n=3 gives an ice sheet with a steeper edge than using n=2, whereas the mean thickness strongly depends on the flow parameter A.

However, the parameter A is not a constant but depends on the properties of ice, in particular on its temperature. From laboratory experiments it is known that the value of A varies with temperature according to

$$(6.1.2) \qquad A = A_0 \, e^{-Q/RT} \, ,$$

where R is the gas constant (8.314 J/mol K) and Q the activation energy for creep. Values of Q, measured at temperatures below $-10\ ^\circ C$, range from 42 to 84 kJ/mol, see for instance Paterson (1981), p. 28; Weertman (1973), table 2. For higher temperatures the value of Q increases with temperature. The coefficient A_0

95

depends on the crystal structure of the ice and on concentrations
of impurities. In table 6.1 (from: Paterson (1981), table 3.3)
the commonly used values of the flow parameter for different
temperatures are given.

We can now ask whether strain heating can raise the ice
temperature so much that the feedback on A leads to a runaway
increase in deformation rate. This phenomena is called creep
instability, which has been investigated by, among others,
Clarke et al. (1977). We briefly summarize their analysis below.

Clarke et al. considered a parallel-sided slab of thickness
H resting on a bedrock with constant slope α_s. Neglecting
horizontal advection of heat, the temperature equation becomes

$$(6.1.3) \qquad \frac{\partial T}{\partial t} = \frac{k}{\rho c} \frac{\partial^2 T}{\partial z^2} - w \frac{\partial T}{\partial z} + \Phi.$$

The dissipative heating Φ is considered to result either from
shear stresses or from longitudinal stresses. In the first case
Φ is a function of the ice thickness H, the surface slope α_s
(H and α_s determine the shear stress) and the temperature T. In
the second case Φ is a function of H, T and the longitudinal
stress τ_{xx}. Introducing a stability parameter β being proportional
to the ratio of deformational heat production to the rate at
which this heat is conducted to the boundaries of the slab, and
using (5.2.2) for the vertical velocity w, Clarke et al. find a
dimensionless equation for the ice temperature θ. Using numerical
procedures this equation can be solved for the stationary case
($\partial \theta / \partial t = 0$). The behaviour of their model is described by the
cusp catastrophe (compare Figure 2.2).

T (°C)	A $(s^{-1} kPa^{-3})$
0	5.3×10^{-15}
-5	1.7
-10	5.2×10^{-16}
-15	3.1
-20	1.8
-25	1.0
-30	5.4×10^{-17}
-35	2.9
-40	1.5
-45	7.7×10^{-18}
-50	3.8

Table 6.1. Values of A for different temperatures (n=3).

In the case of no strain heating ($\beta=0$) the temperature
profiles found by Clarke et al. are the same as the ones given in
section 5.2. When there is no accumulation or ablation (M=0, no
vertical advection) they find that for certain values of β three
steady-state temperature profiles are possible. In Figure 6.1
(after Clarke et al.,1977) the dimensionless basal temperature
is given as a function of the stability parameter β and the
geothermal heat flux. For small values of β only a cold ice sheet
can exist, while for large β the stable solutions represent warm
ice sheets. For intermediate values of β three steady state
solutions appear of which the middle one is unstable. Including
vertical advection does not change the qualitative results given
in Figure 6.1 but causes a shift of the bifurcation points along
the β-axis. Accumulation (resulting in downward advection) tends
to reduce the range of β for which multiple steady states occur,
while upward advection (ablation) increases this range. Clarke
et al. suggest that East Antarctica may be unstable (that is, the
ice sheet is presently going from the cold mode to the warm mode
with higher basal temperatures) and that unstable conditions may
have existed in the central part of ice-age ice sheets.

A few critical remarks should be made concerning the analysis
presented by Clarke et al. One shortcoming is the use of the
temperature equation (6.1.3) which restricts the analysis to the
parts of an ice sheet where horizontal advection is negligible,
i.e. near the ice divide. As long as the horizontal temperature-
gradient is small, this approximation remains valid. But when an
instability develops the ice temperature will increase (or

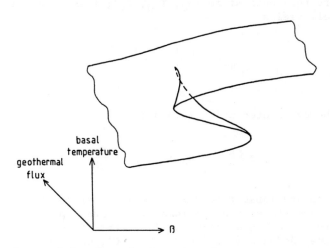

Figure 6.1. Dimensionless basal temperature as a
function of the stability parameter and the geothermal
heat flux. After Clarke et al. (1977).

decrease) locally, thus creating a horizontal inhomogeneity. In
that case the advection of heat becomes important and equation
(6.1.3) cannot be used anymore.

The second shortcoming of the model is the use of the stability
parameter β as control parameter. Any change in the flow parameter
will lead to a change in ice-mass discharge, and thus to a change
in the surface slope and stress distribution. Since β is a
function of the ice thickness H and the surface slope α_s (or the
longitudinal stress τ_{xx} depending on which case is under
consideration), this means that the value of β will change during
transition from the cold to the warm regime (or vice versa).
Therefore, to get a more realistic description of creep
instability it is necessary to include ice dynamics.

In the next section we will investigate this point by
employing the simple model developed in section 5.3.

6.2   The basic mechanism of instability

We consider a zero-dimensional model of a drainage basin of size
L. The rate of change of the characteristic ice thickness H is
given by (8.3.3)

$$(6.2.1) \qquad \frac{dH}{dt} = -C\,\frac{H^{2m+1}}{L^{m+1}} + M.$$

Here C and m are generalized flow parameters. Defining a
stationary reference state (H_o, T_{eo}, C_o), this equation
immediately yields

$$(6.2.2) \qquad H_o = M^{\frac{1}{2m+1}}\,L^{\frac{m+1}{2m+1}}\,C_o^{\frac{-1}{2m+1}}.$$

T_{eo} can be calculated using (5.3.5):

$$(6.2.3) \qquad T_{eo} = [\frac{g}{c} - \gamma]\,H_o + T_{air} + \frac{G}{\rho cM}.$$

The temperature lapse rate along the ice surface is denoted by γ.
The reference state is determined by the mass balance M and the
half-width L (both quantities will be kept fixed). To include the
temperature feedback we write $C = C_o + C'$, where C' is a function
of temperature. Similarly, we set $H = H_o + H'$ and $T_e = T_{eo} + T_e'$.
Using (6.2.2) yields

(6.2.4) $H' = [\frac{\partial H}{\partial C}]_o C' = - \mu C'$,

where $\mu = \frac{1}{2m+1} \frac{H_o}{C_o}$.

The temperature T'_e immediately follows from (5.3.5) by inserting the expression for T_e given above and subtracting the reference temperature. We find

(6.2.5) $T'_e = - \Gamma H'$,

with $\Gamma = - [\frac{g}{c} - \gamma]$.

At this point something should be said about the temperature-dependence of the flow parameter C'. The only temperature we can calculate is the temperature of the ice leaving the sheet at the edge. It seems appropriate to make C' a function of T'_e, because the basal temperature of an ice sheet (which essentially governs the ice-mass discharge) is likely to be directly related to T_e. Therefore we write

(6.2.6) $C' = k \arctan(\nu T'_e)$.

So we assume that the flow parameter increases with temperature, but is nevertheless bounded. This simple form is chosen to facilitate the analysis. For small values of $\nu T'_e$ (6.2.6) can be approximated as

(6.2.7) $C' = k[\nu T'_e - \frac{\nu^3}{3} T'^3_e]$.

Substituting (6.2.5) and (6.2.7) in (6.2.4) yields

(6.2.8) $H'^3 + aH' = 0$

where the relevant control parameter apparently is

$$a = \frac{\frac{1}{\mu} - \nu\Gamma k}{\frac{1}{3}\nu^3 k\Gamma^3} .$$

Equation (6.2.8) has only three solutions when a is negative; when a is positive only the reference state H'=0 represents a real solution (see Figure 6.2). Now ν, k and μ are always positive while Γ has to be positive to yield realistic results, so multiple equilibria occur when

(6.2.9) $\Gamma > (\mu\nu k)^{-1}$.

    According to this analysis distinct flow regimes can occur provided that the lapse rate is larger than g/c (=4.9 K/km) and condition (6.2.9) is fulfilled. The possibility of three steady states increases with increasing strength of the temperature – ice flow feedback (ν), increasing dependency of the ice thickness on the flow parameter (μ) and increasing lapse rate in the atmosphere. So the dependence of the flow parameter C on the ice temperature provides one mechanism for instability (or better: multiple steady states).
    When writing down an equation of the type (6.2.6) it is of course more or less assumed that T_{eo} is close to the melting point, because here the largest changes in the flow parameter C

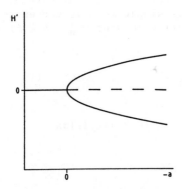

Figure 6.2. Steady-state solutions for a zero-dimensional ice-sheet model as a function of the control parameter a. The dashed line represents the unstable steady state (scales are chosen arbitrarily).

are expected. A more explicit approach is to make C a function
of the rate of melting at the base. We write

(6.2.10) $C' = C_1 \arctan(C_2 \, dW/dt)$ if $dW/dt > 0$,

                  $C' = 0$ if $dW/dt < 0$.

The production of melt water immediately follows from (5.3.5):

(6.2.11) $\dfrac{dW}{dt} = \dfrac{c}{L} \, M \, (T_e - T_m)$,

where L is the latent heat of fusion (335 kJ/kg) and T_m the melt
temperature. When dW/dt has been calculated, T_e must of course
be set to T_m. The choice of C_2 does not have too much influence
on the bifurcation properties, so a constant value will further
be used (10^4yr/m).
     The set of equations (6.2.1), (5.3.5) and (6.2.10) cannot be
solved analytically. So to find the steady states, (6.2.1) has
to be integrated numerically, each time step using (5.3.5) to
calculate T_e, using (6.2.11) to see whether melt water is formed,
and using (6.2.10) to calculate the flow parameter. Figure 6.3
shows the solution for $T_{air} - T_m = 3.5$ K, $\gamma = 9$ K/km, M = 0.2 m/yr.

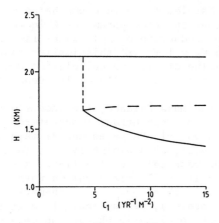

Figure 6.3. Steady-state ice thickness as a function of
the flow parameter C_1. The values of the other parameters
are: $C_o = 1$ m^2 yr^{-1}, $C_2 = 10^4$ yr m^{-1}, L = 1000 km. The
dashed line gives the unstable steady state. From
Van der Veen and Oerlemans (1984).

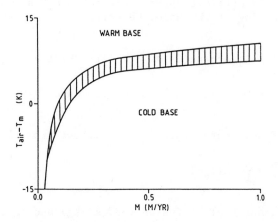

Figure 6.4. Regimes of a polar ice sheet in dependence
of the environmental conditions (mass balance and
atmospheric temperature). The shaded area shows the
region where both warm and cold-base states can be
in equilibrium. From Van der Veen and Oerlemans (1984).

The unstable equilibria were obtained by integrating with a
negative time step.
    Apparently, the slow mode, without melt water present, is
always a steady state. Only for values of C_1 larger than $4 \ m^{-2} \ yr^{-1}$
three steady states are possible, the middle one being unstable.
It is generally believed that the flow parameter C increases by
an order of magnitude when production of melt water is substantial
e.g. Weertman (1969). With $C_0 = 1 \ m^{-2} \ yr^{-1}$ this implies that
$C_1 = 7 \ m^{-2} \ yr^{-1}$. With this value of C_1, the calculation was
repeated for other values of the mass balance. The results are
summarized in Figure 6.4.
    For very low values of the mass balance, ice sheets with a
cold base are not possible. For sea-surface temperatures
prevailing at Antarctica (about -15 $^{\circ}$C as an annual mean value),
the flow regime of an ice sheet, or drainage system, turns out
to be very sensitive to the mass balance. The figure also shows
that the region where multiple equilibria occur is rather small.
Although the model employed in this section is schematic (but
exact in terms of the heat budget), we can learn from it that
the reaction of an ice sheet to climatic change can very well
be steered by variations in the mass balance. This point is
relevant, for instance, when one studies the response of the
Antarctic Ice Sheet to a developing ice-age climate.
    To supplement the qualitative study of this section, we now
consider more explicit calculations with a numerical model.

6.3 Numerical experiments with a temperature dependent flow law

The approximate method to calculate the temperature field
developed in section 5.4 can be used to study the interaction
between ice flow and temperature field. This requires the flow
parameter in the flow law to be related to temperature. We know
that generally most of the velocity shear takes place in the
lower ice layers, so it is natural to express the flow constant
in the basal temperature Θ_0. We assume that the temperature
dependence is similar to that for local deformation. Denoting
ice temperature with respect to the melting point by Θ_0', we use

$$(6.3.1) \qquad B = B_0 \, \exp\{-16750/(273.15+\Theta_0')\} \; ,$$

where Θ_0' is in $^\circ$C.

To investigate the implications of a variable flow parameter
B, we consider a two-dimensional ice sheet of fixed size, subject
to an ice-accumulation rate independent of location. The ice flow
is calculated with the scheme described in section 4.1, combined
with the thermodynamic calculation outlined in section 5.4. The
half-width of the ice sheet is set to 1500 km, covered by 31 grid
points. In the centre of the ice sheet all horizontal derivatives
are set to zero, while at the edge the boundary condition is
H=0. Since grounding-line dynamics are not included, a simple
damped return to local isostatic equilibrium is included according
to

$$(6.3.2) \qquad \frac{\partial b}{\partial t} = - \, (b + H/3.2)/T^{*} \; .$$

Here b is the bedrock elevation (initially b(x)=0), and the time
scale T^{*} for isostatic adjustment is set to 5000 yr. Other model
constants are: γ=-10 K/km (atmospheric lapse rate), B_0=2.3x10^{26}
m$^{-3/2}$yr^{-1}, n=2.5, ψ=0.042 W/m (geothermal heat flux).

To obtain a picture of the basic sensitivity of the ice sheet,
a series of runs was performed in which sea-level temperature and
mass balance M were systematically varied. Each integration was
continued until a steady state appeared to be reached. This takes
about 30 000 to 70 000 years of simulated time, depending on the
mass balance.

Figure 6.5 summarizes the dependence on sea-level temperature
in terms of mean ice thickness $\bar{H}$ and mean basal ice temperature
$\bar{\Theta}_0$. The shape of the model ice sheet and some temperature
profiles for sea-level temperatures of -16 $^\circ$C and -12 $^\circ$C are
shown in Figure 6.6. All these results apply to a mass balance of
0.2 m ice depth / yr. It is clear from Figure 6.5 that the ice
thickness decreases substantially when sea-level temperature

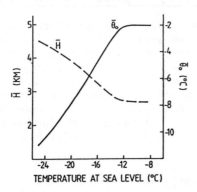

Figure 6.5. Mean ice thickness and basal temperature of a polar ice sheet as a function of sea-level temperature.

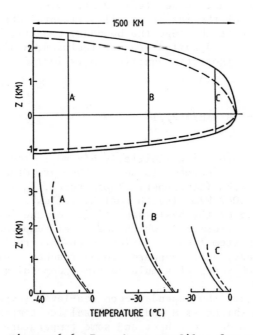

Figure 6.6. Ice-sheet profiles for sea-level temperatures of -16 °C (solid line) and -12 °C (dashed line). The lower part of the figure shows corresponding temperature profiles.

rises. For a sea-level temperature of -12 °C the base is at the
melting point almost everywhere, and, since the effect of basal
melting on the ice-mass discharge is not yet taken into account,
a further increase in atmospheric temperature does not have any
effect. In the next section we will deal with basal melt water.

It can be seen in Figure 6.6 that, apart from a difference
in mean ice thickness, a lower air temperature leads to a
considerably steeper ice-sheet edge. The reason is that for a
cold ice sheet (no basal melting) the lowest basal temperature,
and thus the smallest flow parameter, is found near the edge. It
is also interesting to note that a reversed temperature gradient
in the upper ice layers only occurs for warmer conditions
(dashed lines in the figure). This probably is a consequence of
the larger temperature gradient along the ice surface. Other
experiments brought to light that also larger accumulation
rates tend to produce stronger temperature inversions.

In Figure 6.7 $\bar{H}$ and $\bar{\Theta}_0$ are shown as a function of the mass
balance M, for a sea-level temperature of -16 °C. The result very
clearly shows how important M is with regard to the mean basal
temperature. In particular for lower values of M the sensitivity
is large. The mean ice thickness curve shows that an increase in
M from 0.1 to 0.3 m ice depth / yr leads to an increase in $\bar{H}$ of
1300 m ! This large change in ice thickness contains a
substantial contribution from the temperature dependence of the
flow law. The cooling of the ice sheet associated with an
increase in M causes further growth of the ice volume. To further
illustrate this point, the dependence of $\bar{H}$ on M for a fixed flow
parameter B is shown by the thin dashed line. Here B was chosen
in such a way that the constant B and Θ_0-dependent B give the
same ice thickness for M=0.2 m ice depth / yr. So the conclusion

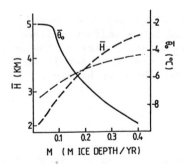

Figure 6.7. Ice thickness and basal temperature in
dependence of the mass balance M. The thin dashed line
shows the result for a fixed flow parameter (no
temperature feedback on the ice flow).

obviously should be that the temperature dependence of the flow
law approximately doubles the sensitivity of the ice sheet to
changes in the accumulation rate.

A natural question to ask now is whether local variations in
the mass balance are important. Generally, M is largest near the
edge of the ice sheet and decreases towards the centre. The
present model enables us to make an interesting comparison.
Figure 6.8 shows calculated basal temperatures for the (more
realistic) case in which a maximum in the mass balance is found
near the ice-sheet edge (solid lines), and in case of a uniform
mass balance equal to the average mass balance in the other case
(dashed lines). Apparently larger values of M near the edge lead
to higher basal temperatures, and consequently to lower mean
ice thickness (the values for $\overline{H}$ are 2826 and 3273 m). The
differences found here are due to the large frictional heating
near the edge in case of the large mass balance.

To conclude this section we mention that in the computations
no trace was found of multiple steady-state solutions. The
temperature dependence of the flow parameter appears to be too
weak to create bifurcation in the way described qualitatively
in section 6.2.

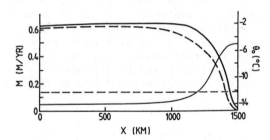

Figure 6.8. Comparison of steady state basal
temperatures calculated for a uniform (dashed lines) and
x-dependent mass balance (solid lines). The prescribed
sea-level temperature is -16 °C. Heavy lines give
temperatures, thin lines the mass balance.

6.4. The effect of basal water

Production of melt water at the base of an ice sheet or glacier
and its effect on the ice-mass discharge is a poorly understood
process. For small mountain glaciers theories have been developed
that seem to be supported to some extend by scarce observations
(for reviews see Weertman, 1972 and Lliboutry, 1979).

When the pressure melting point is reached and the heat budget

is still positive melt water will be produced. If a sufficiently
large area is involved the ice mass may start to slide over the
bed, and additional frictional heating will then lead to enhanced
melt-water production. This process thus is of a runaway
character. The ultimate steady state (if existing) is, among
other factors, determined by how basal water is removed. In case
of a comparatively steep bedrock, chanelling generally occurs.
This situation is normally encountered when mountain glaciers are
investigated.

How basal water under a large continental ice sheet on a
relatively flat bedrock spreads is a different matter.
Temperature and permeability to water of the bedrock surrounding
a region where basal melting occurs will without doubt be
important factors. When basal water cannot be removed at a
sufficiently high rate, subglacial lakes may form. Radio echo
soundings have indicated that on Antarctica such subglacial
water bodies indeed exist (e.g. Drewry, 1983), but definite
evidence is still not available.

Weertman studied the implications of the presence of melt
water on the dimensions of large ice sheets. His analysis
indicates that the mean ice thickness can be halved in case of
sufficient melt-water production. Weertman and Birchfield (1982)
recently applied the same basic ideas to ice streams, in order to
study the stability of the West Antarctic Ice Sheet.

At present a reliable scheme suitable for incorporation into
time dependent numerical models does not exist. Nevertheless we
can investigate the potential importance of basal water by
including the effect in a very schematic way. The simplest
approach to deal with flow of basal water is to use a diffusive
equation, i.e.

(6.4.1) $$\frac{\partial W}{\partial t} = - D_w \nabla^2 W + S .$$

Here W is the amount of basal water, D_w the diffusivity and S
the rate of basal melting (when freezing occurs, S<0).

The next step is to relate the ice-mass discharge to the
amount of basal water. We could for example set

(6.4.2) $$B = B_o + B_1 W .$$

So the flow parameter now contains a contribution from internal
deformation (B_o) and sliding $(B_1 W)$. Since we now treat the effect
of basal water in a speculative manner anyway, we ignore the fact
that velocity profiles will change. The only modification we make
to the previous model is including (6.4.2) and use all frictional
heating associated with the $B_1 W$ contribution for additional

melting. The basic positive feedback is then included.

A new series of numerical experiments with the effect of basal water included now yields Figure 6.9. By starting from different initial conditions, multiple steady states are now found for a range of sea-level temperatures. The results shown are for M=0.2 m ice depth / yr. Note the qualitative agreement with Figure 6.4. Although not calculated here, a plot of $\overline{H}$ versus M for a given sea-level temperature will give a similar result. According to Figures 6.5 and 6.7, the effects of increasing sea-level temperature and decreasing mass balance are interchangeable to some extent.

For the fast-flow mode corresponding to a sea-level temperature of −20 °C and a mass balance of 0.2 m ice depth / yr, the steady state profile, amount of basal water and rate of basal melting are shown in Figure 6.10. When looking at this picture it is once more important to realize that the effect of basal water is treated in a very schematic way. The thickness of the water layer should be considered as indicating the total amount of water present in a grid square rather than the physical thickness of a uniform layer of water.

In the previous calculations the value used for D_w was 10^9 m^2/yr. This value was arbitrarily chosen, and of course determines the average amount of basal water found when an ice sheet is in the 'fast mode'. What really counts with respect to the dynamics of the ice sheet is the difference in flow parameter between the basal water and no-basal water cases, so the relevant quantity

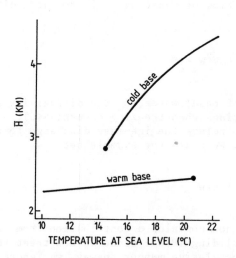

Figure 6.9. Stable steady-state ice thickness as a function of sea-level temperature. The mass balance is 0.2 m ice depth / yr.

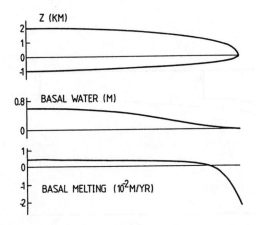

Figure 6.10. Characteristics of the steady-state ice
sheet corresponding to a sea-level temperature of
-20 °C and a mass balance of 0.2 m ice depth / yr.

seems to be D_W/B_1. It should be expected that other approaches
to modelling the flow of basal water yield different results.
Such possibilities are for instance making D_W a function of
hydrostatic pressure, or advecting the basal water with a fraction
of the mean ice velocity. In the latter case two distinct types
of flow still show up, but the critical points in Figure 6.9
come much closer (Oerlemans, 1983).

6.5 The effect of temperature-dependent flow on the equilibrium profile of a free floating ice shelf

In section 3.9 expressions were given for the steady-state
profile of an unconfined ice shelf. As was shown in Figure 3.8,
the profile of an ice shelf strongly depends on the mass balance
M. Basal melting reduces the size of the shelf considerably.
However, not only the mass balance, but also the flow parameter
A influences the thickness of the shelf. This follows immediately
from equation (3.8.9)

$$(6.5.1) \qquad Hu = Mx + H_o u_o \ .$$

Since the ice velocity increases with A, the ice thickness
becomes smaller when the temperature of the ice shelf increases.
This implies that the shelf will break off sooner (keeping the

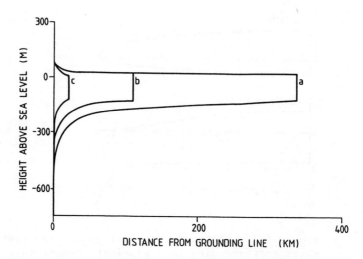

Figure 6.11. Effect of temperature on the steady-state ice-shelf profile. Constants used are: H_o= 800 m, u_o= 100 m/yr, H_{cr}= 150 m, M = 0, A = 10^{-25} (profile a), A = 3.1x10^{-25} (profile b) and A = 17.0x10^{-25} $N^{-3}m^6s^{-1}$ (profile c).

ice flux across the grounding line constant). In Figure 6.11 this effect is illustrated. Three ice-shelf profiles are shown with a mutual temperature difference of approximately 10 °C. In calculating these profiles the mass balance was kept fixed, which is a rather artificial situation since basal melting or freezing will of course depend on the ice temperature.

From the examples given in this section and in section 3.9 we may conclude that a free floating ice shelf can only exist when its temperature is not too high and basal melting rates are low. The temperature of an ice shelf is influenced by the temperature of the ice flowing across the grounding line and by the air temperature. The basal temperature is fixed at the freezing point of sea water. So the flow parameter should not be related to the basal temperature (which gives a good description of the temperature – ice flow feedback in an ice sheet, as shown in section 6.3) but to for example the vertical mean temperature. This seems a good approximation regarding calculations done by Sanderson and Doake (1979). They found that in an ice shelf vertical shear is negligible, compared to the normal stress gradient, even when the vertical temperature profile is taken into account. This means that differential flow (that is, the difference in velocity at the surface and the base of the shelf) is negligible. Therefore, taking a constant value for A, related to the mean temperature, will lead to realistic results.

7. BEDROCK ADJUSTMENT

As we have seen in the previous chapters an ice sheet can
easily become a few kilometers thick. For example, the ice
thickness on Antarctica reaches values of more than 4500 m
with a mean value of about 2000 m. So a considerable pressure
is exerted on the underlying bedrock, which will deflect to
restore the equilibrium of forces. In some ice-sheet models
the effect of ice thickness on bedrock topography is
incorporated by assuming local isostatic adjustment (with a
time lag of 4000 years, say).However, the rigidity of the
lithosphere gives rise to deviations from isostatic equilibrium,
especially near the edges of the sheet.
   In this chapter we will investigate how the adjustment of
the bedrock to an ice load can be calculated. First the
deflection of the lithosphere, which gives the ultimate
depression of the bedrock, is treated. The response time, i.e.
the time it takes for the bedrock to reach a new equilibrium,
is governed by the flow in the asthenosphere. We will turn to
that in section 7.3.

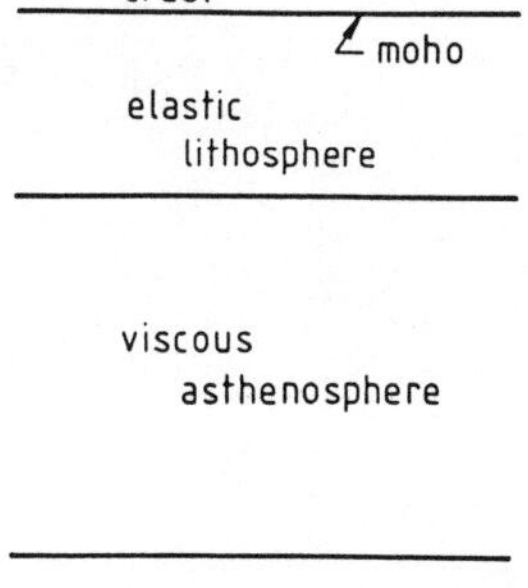

Figure 7.1. Schematic picture of the outer shell of
the solid earth. We treat the upper part as an elastic
plate, and the lower part as viscous material.

7.1. Introduction

Since many readers of this book will not have a background in
geology, we first give a brief overview of the structure of the
crust. Figure 7.1 shows how the outer shell of the earth is
built up. Starting at the surface we first find the crust made
up of relatively light rock. A typical thickness for the
oceanic crust is 6 km and for the continental crust 35 km.
The boundary between the crust and the underlying mantle is
usually well-defined. However, the mechanical properties of
the crust and the underlying lithosphere can be considered to
be similar, at least for the present purpose. So we will make no
distinction between crust and lithosphere, and refer to the
lithosphere as being the outer part of the solid earth.
    The lower boundary of the lithosphere is usually defined by
an isotherm, mostly the 1600 K-isotherm. Rocks above this
isotherm do not significantly deform while rocks beneath this
isotherm are so hot that solid-state creep can occur. This
lower, warmer layer of rocks is called the asthenosphere.
The adjustment of the bedrock to the load of an ice sheet takes
place in the lithosphere and the asthenosphere.
    The first ice-sheet models incorporating bedrock adjustment
made use of isostatic equilibrium. This is illustrated in
Figure 7.2. According to this principle the lithosphere is in
floating equilibrium with the underlying heavier substrate of
the astenosphere. In other words the total mass in a vertical
column is constant. So, when a mountain is present, the
lithosphere will bend downward to compensate for the excess of

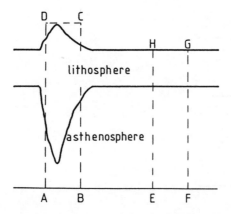

Figure 7.2. Illustration of isostatic equilibrium. The
lithosphere under a surface load bends downward in such
a way that the total mass in a column remains constant.

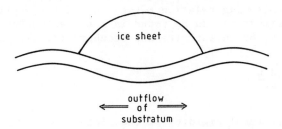

Figure 7.3. Flow in the asthenosphere due to a growing
ice sheet.

mass (remember that the lithosphere is made up of relatively
light rock). Referring to Figure 7.2, we see that the
lithosphere is deflected under the mountain so that the total
mass in column ABCD equals the mass in column EFGH (neglecting
air density). What happens when an ice sheet forms is that the
lithosphere bends downward until an equilibrium is reached.
The deflection of the lithosphere immediately follows from the
ice thickness H: $w = \rho/\rho_m H$ (ρ is ice density, ρ_m is mantle
density).

   Because of the viscous behaviour of the asthenosphere a new
isostatic equilibrium will be reached only after thousands of
years. The outflowing substratum causes the surrounding land
to rise, see Figure 7.3. Subsequently, when the ice sheet melts
down the part of the lithosphere once covered by ice will rise
again and the surrounding parts will sink.

   Although the theory given above is rather simple, it is able
to explain at least qualitatively the observed features of
bedrock adjustment. However, if we want to model this more
properly we have to consider the lithosphere and the
asthenosphere in more detail. Because of the rigidity of the
lithosphere the ultimate depression of the bedrock is
determined by the lithosphere. The properties of the viscous
asthenosphere govern the rate of bedrock adjustment. In the
following sections we will discuss how adjustment of the
bedrock to an ice load can be calculated and incorporated in
a numerical ice-sheet model.

7.2   Deflection of the lithosphere

As mentioned in the previous section, the lithosphere remains
rigid on geological time scales. This means that flexure of
the lithosphere is similar to bending of a rigid elastic plate

(see Figure 7.4). We denote the deflection by w(x), positive downwards. The following relation between w and the applied load q(x) holds (a derivation can be found in most textbooks on elasticity theory, or in Turcotte and Schubert (1982), chapter 3):

$$(7.2.1) \qquad D \frac{d^4w}{dx^4} = q(x) \quad ,$$

where D is the flexural rigidity of the plate.

In case of deflection of the lithosphere due to the presence of an ice sheet, we should include in q(x) the buoancy force that tries to restore the undisturbed situation. Denoting the density of the rock underlying the lithosphere by ρ_m, this upward force equals $gw\rho_m$. If $q_a(x)$ is the applied load at the surface (the weight of the ice sheet), we can write

$$(7.2.2) \qquad D \frac{d^4w}{dx^4} + \rho_m gw = q_a(x) \quad .$$

Now consider the case in which a line load V_i is applied at x=0 (as shown in Figure 7.4). The solution to (7.2.2) then reads (see e.g. Turcotte and Schubert, 1982)

$$(7.2.3) \qquad w(x) = \frac{1}{8D} V_i \alpha^3 e^{-x'} \{\cos x' + \sin x'\} \quad ,$$

$$\text{where } x' = |x|/\alpha \quad \text{and} \quad \alpha = (4D/\rho_m g)^{1/4} \quad .$$

In Figure 7.5 the solution (7.2.3) is sketched. An interesting feature is the presence of a forebulge at a distance of approximately 3α from the line load. Such a forebulge, resulting from the rigidity of the lithosphere, can for example be found

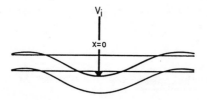

Figure 7.4. Bending of a rigid elastic plate under a line load.

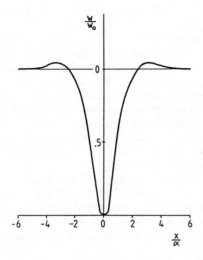

Figure 7.5. Deflection of the lithosphere as given
by (7.2.3).

in the vicinity of the Hawaiian Archipelago.

Observations suggest that the value of α is about 180 km
(Walcott, 1970). With ρ_m = 3300 km/m^3 this gives for the flexural
rigidity of the lithosphere: D = 8.5x10^{24} N m. It turns out that
the choice of D does not affect the actual deflection very much.
The position of the forebulge, however, is influenced by the
flexural parameter α.

We now turn to the problem of how the deflection can be
calculated for an arbitrary ice sheet. We consider a typical
'model ice sheet', i.e. we suppose that values of the ice
thickness are known on a grid. Since (7.2.2) is linear, the
total deflection of the lithosphere can be written as the sum
of deflections caused by line loads at x=0, x=δx, x=2δx, etc.
In the grid point where grounded ice occurs the load is

$$(7.2.4) \qquad V_i = \rho g \, H_i \, \delta x \quad .$$

However, when the bedrock is below sea level we have to take into
account the weight of water not present in the undisturbed
situation. Thus in grid points without grounded ice the load
should be written as

$$(7.2.5) \qquad V_i = \rho_w g (h_b - h_{bo}) \, \delta x \quad ,$$

where ρ_w is the density of sea water, h_{bo} the undisturbed and h_b the actual bedrock profile.

In Figure 7.6 the depression of the lithosphere under the weight of an ice sheet as calculated with the procedure outlined above is sketched. Also shown is the deflection in case of pure isostatic equilibrium ($w_{iso} = H \rho/\rho_m$). The differences between the curves are generally small, except near the edges of the ice sheet. Since the bedrock topography near the ice-sheet edge determines to some extent whether an ice sheet will retreat or advance, in particular when the bedrock is below sea level, it may be necessary to include the rigidity of the lithosphere in the calculation of bedrock adjustment.

Again, the forebulge appears, although its height is rather small, namely, only 4 % of the depression under the central part of the ice sheet. For a sheet with a maximum thickness of 3 km, as used in the computation of Figure 7.6 , the height of the forebulge is thus approximately 20 m. This is much smaller than the values estimated to have occurred in North America, being 80 m. To explain a forebulge of this size by elastic bending of the lithosphere would require a 8 km thick ice sheet, which is obviously unrealistic. Here transient effects presumably play a role.

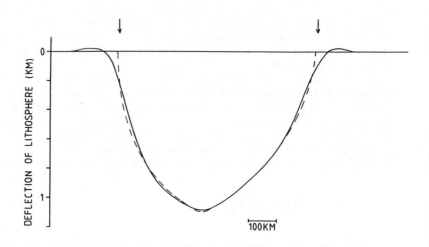

Figure 7.6. Deflection of the lithosphere due to the presence of an ice sheet, for local isostatic adjustment (---), and taking into account the rigidity of the lithosphere (——). Arrows indicated the edges of the ice sheet. Grid distance used: 25 km.

7.3 The flow in the asthenosphere

We assume that the asthenosphere behaves like a viscous fluid.
For the case under consideration, the pressure exerted by the
lithosphere results in an approximately horizontal flow in the
asthenosphere. We may safely neglect accelerations, implying that,
when horizontal gradients in the velocity u are small, the
equation of motion becomes

$$ - \frac{1}{\rho_m} \frac{\partial(p-p_0)}{\partial x} + \nu \frac{\partial^2 u}{\partial z^2} = 0 \quad , $$

or, alternatively,

$$ (7.3.1) \qquad \frac{\partial^2 u}{\partial z^2} = \frac{1}{\nu \rho_m} \frac{\partial(p-p_0)}{\partial x} \quad . $$

Here ν denotes the viscosity of the asthenosphere and p_0 the
hydrostatic pressure prevailing when there is no motion. The
geometry is shown in Figure 7.7. We take z=0 as reference level,

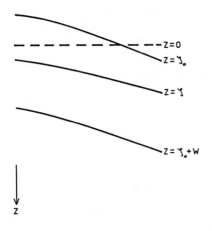

Figure 7.7. Geometry used in the discussion on the flow
in the asthenosphere. Note that we now take z positive
in downward direction !

$z=\zeta_0$ denotes the undisturbed upper boundary of the asthenosphere, and z=H its lower boundary (which is in fact a rather artificial boundary used in the present derivation). The actual upper boundary of the asthenosphere is at $z=\zeta$. As boundary conditions we have for the horizontal velocity field

$$u = 0 \quad \text{at} \quad z = \zeta \quad ,$$

$$\partial u/\partial z = 0 \quad \text{at} \quad z = H \quad .$$

The last condition merely states that the asthenosphere exerts no stress on the underlying material.

Since changes in the thickness of the asthenosphere are very small compared to its thickness (the ratio is about 1 to 100), we may replace the first boundary condition by

$$u = 0 \quad \text{at} \quad z = 0 \quad .$$

Integrating (7.3.1) twice with respect to z yields, after inserting the boundary conditions

(7.3.2) $$u(z) = \frac{1}{2\nu \, \rho_m} \frac{\partial(p-p_0)}{\partial x} (z^2 - 2Hz) \quad .$$

When the asthenosphere has adjusted itself completely to the ice load, the flow will cease. Now changes in pressure are only due to changes in ζ, so we may write

(7.3.3) $$p - p_0 = \rho_m g \, (\zeta_0 - \zeta + w) \quad .$$

Here the term $\rho_m g w$ stems from the pressure exerted by the ice load and the rigidity of the lithosphere. When there is no ice sheet the asthenosphere will keep its original thickness $H-\zeta_0$, while the effect of an ice sheet thus is to thin the asthenosphere until its thickness is $H-(\zeta_0+w)$. From (7.3.2) and (7.3.3) we acquire

(7.3.4) $$u(z) = \frac{1}{2} \frac{g}{\nu} (z^2 - 2Hz) \frac{\partial}{\partial x}(\zeta_0 - \zeta + w) \quad .$$

Now consider a vertical column extending from z=H to $z=\zeta$, with a cross section of unit area. Conservation of mass requires

(7.3.5) $$\frac{\partial \zeta}{\partial t} = - \frac{\partial}{\partial x} \int_{\zeta}^{H} u \, dz \quad .$$

We approximate the integral as

$$(7.3.6) \qquad \int_{\zeta}^{H} u\,dz \simeq \int_{0}^{H} u\,dz = -\frac{gH^3}{3\nu}\frac{\partial}{\partial x}(\zeta_0 - \zeta + w) \quad,$$

where (7.3.4) has been used for the velocity profile. Combining (7.3.5) and (7.3.6) we now obtain an expression for the time-dependent behaviour of the thickness of the asthenosphere. It reads

$$(7.3.7) \qquad \frac{\partial\zeta}{\partial t} = \frac{gH^3}{3\nu}\frac{\partial^2}{\partial x^2}(\zeta_0 - \zeta + w) \quad.$$

Since the thickness of the asthenosphere directly determines the depression of the bedrock, we are finally arriving at an equation for the rate of change of bedrock elevation. Denoting the undisturbed bedrock profile by h_{bo} (positive when below sea level), we obtain

$$(7.3.8) \qquad \frac{\partial h_b}{\partial t} = \frac{gH^3}{3\nu}\frac{\partial^2}{\partial x^2}(h_{bo} - h_b + w) \quad.$$

This is a diffusion equation, and the characteristic time scale for bedrock sinking thus turns out to increase with the length scale L of the load according to $T = 3\nu L^2/(gH^3)$, see Table 7.1.

$D_a(km^2/yr)$	L (km)	T (yr)
35	500	7143
35	1000	28571
40	500	6250
40	1000	25000
45	500	5556
45	1000	22222
50	500	5000
50	1000	20000

Table 7.1. Relaxation time of bedrock adjustment, for different values of the diffusivity D_a (governing the viscous behaviour of the asthenosphere). L is the typical length scale of the load.

The diffusivity for asthenospheric flow as modelled here apparently is $D_a = gH^3/(3\nu)$. The values of D_a used in the table all are in the range given by Walcott (1973), as calculated from observations in North America and Scandinavia. These observations only give an estimate of D_a, not of ν, because the appropriate value of H is not known, in fact. Values for H given in the literature vary from one hundred to several hundreds of kilometers.

7.4 An example

In this section we compare two integrations with a numerical ice-sheet model, one with and one without a calculation of bedrock adjustment.

For simplicity we take an intially flat bedrock, and the geometry and mass balance are the same as used in section 4.2, namely,

$$M = 0.5 - 0.5 \times 10^{-6} \, x \quad m/yr \quad .$$

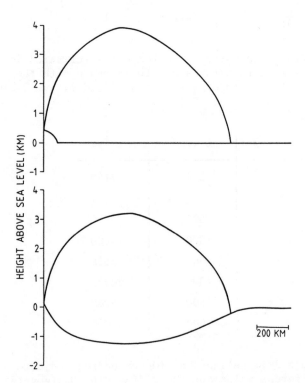

Figure 7.8. Steady-state profiles of an ice sheet on a flat bedrock without (a) and with (b) bedrock sinking.

The spacing between grid points is 25 km, and the integration was started with zero ice thickness everywhere.

Figure 7.8a shows the steady-state profile obtained in case of no bedrock adjustment. Including the latter leads to the profile shown in the lower panel. Here the following parameter values were used: $D = 5 \times 10^{23}$ N m (corresponding to $\alpha = 88.6$ km), and $D_a = 1.67$ m^2/s. According to Figure 7.8b the height of the forebulge appears to be very small. However, a closer look at the transient behaviour reveals an interesting feature. The height of the forebulge reaches a maximum value of 240 m after about 20 000 yr of simulated time, and then decreases towards a value of about 15 m (see Figure 7.9). This behaviour results from the diffusive character of the flow in the asthenosphere, and could explain the height of the forebulges observed in North America. The mechanism is as follows.

When an ice sheet builds up, the load causes the substratum under the ice sheet to flow away in horizontal direction. Since the flow is highly viscous, it takes a rather long time before the surrounding substratum starts to flow. As a consequence, the bedrock near the edges of the ice sheet rises considerably and sinks again when the remote substratum is displaced. Also, when an ice sheet melts down in a short period, the bedrock in the vicinity of the ice-sheet edge will sink a few hundred meters, while the bedrock under the central part of the ice sheet rises immediately. The matter is complicated, of course, by the fact that an ice sheet does not melt uniformly but mainly at its edges.

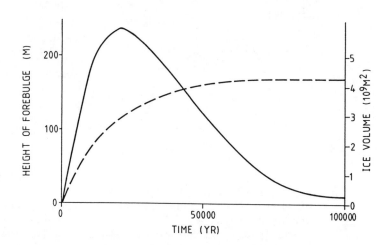

Figure 7.9. Growth and decay of the forebulge (solid line) and ice volume (dashed line) as a function of time.

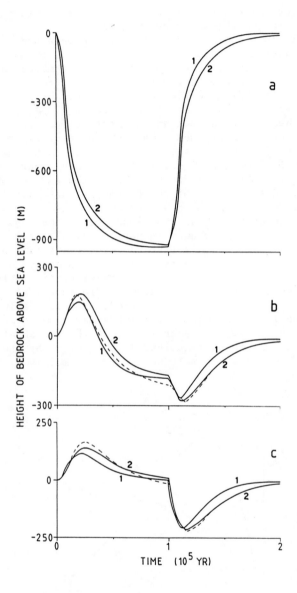

Figure 7.10. Bedrock adjustment under the central part
of an ice sheet (panel a), at the southern edge (panel b)
and at a distance of 900 km from the centre of the ice
sheet (panel c). Parameter values used for the
calculation of bedrock adjustment are: curves 1: D =
5×10^{23} N m, D_a = 1.67 m^2/s; curves 2: D = 5×10^{23} N m,
D_a = 1.13 m^2/s; dashed lines: D = 8.5×10^{24} N m, D_a =
1.13 m^2/s. In a the dashed line coincides with curve 2.

Some results from further numerical experiments are displayed
in Figure 7.10. In these calculations the mass balance was made
strongly negative after 100 000 yr, so that a rapid decay of the
model ice sheet occurred. Panel a shows the bedrock sinking under
the central part of the ice sheet. The curves, which are for
different values of the diffusivity of the asthenosphere, in fact
are a lagged reflection of the ice volume.

In panel b, the bedrock sinking in the last grid point that,
in the steady state, is covered with ice is given. When the ice
sheet is still small, the outflowing substratum causes the
bedrock in this grid point to rise. Later, when the ice sheet
edge comes closer, sinking occurs. In this example the effect
of a rapidly decaying ice sheet is curious: in the grid point
under consideration, the bedrock sinks by about 70 m to compensate
for the inward (i.e. to the decaying central part of the sheet)
flow of asthenospheric rock.

Panel c finally shows the bedrock elevation at the grid point
where, in the steady state at 100 000 yr, the forebulge is found.
Here changes in bedrock elevation are quite similar to those in
the previously discussed grid point.

Comparing the various curves in Figure 7.10, we find that the
results are rather insensitive to the particular choice of D
and D_a. This is not unexpected, because all equations used to
describe the geodynamics are linear. Another implication of this
linearity is that results for the two-dimensional case, which
has not been discussed here, will be rather similar. For two-
dimensional ice-sheet models a diffusion equation for the flow
in the asthenosphere can again be used.

Readers interested in the two-dimensional formulation of flexure,
or in a more general background, may consult Turcotte (1979),
Turcotte and Schubert (1982) and Peltier (1982).
We did not discuss the problem of how global sea level is
related to the evolution and decay of ice sheets. Information
on this can for instance be found in Peltier (1980) and Clark
et al. (1978).

8. BASIC RESPONSE OF ICE SHEETS TO ENVIRONMENTAL CONDITIONS

The mass balance, or annual ice-accumulation rate, of an ice sheet is the net resultant of a number of physical processes (snowfall, freezing of water, evaporation, etc.), and the contribution of a particular process may vary considerably from place to place. Nevertheless, to study the basic response of continental ice sheets to changes in environmental conditions, it is sufficient to express the mass balance in terms of height with respect to the snow line. At the snow line the mass balance is zero by definition. For schematic calculations it is sufficiently accurate to set the mass balance M linearly proportional to $z - h_s$, where z is elevation (above sea level) and h_s is the height of the snow line.

A further discussion of processes making up or affecting the mass balance will be given in the next chapter. For the simple examples discussed below, a sophisticated parameterization of M is not needed.

8.1 Ice sheets on a bounded continent

Apart from the climatic conditions, the geometry of the continent on which an ice sheet might eventually appear is very important. An ice sheet will not be able to grow beyond the edge of the continental shelf (where the ocean depth typically increases from 200 to 3000 m, or so), because the ice starts to float and breaks off to form ice bergs. So the edge of the continental shelf forms a natural bound for a continental ice sheet.

Let us consider a bounded continent and suppose that the snow-line elevation h_s is constant (i.e. independent of location). As pointed out earlier the profile of a perfectly plastic ice sheet only depends on the yield stress and the width of the sheet, not on the mass balance as long as it is positive. This makes it very easy to find out how an ice sheet on a bounded continent reacts to changes in h_s.

For simplicity we assume that the continent is flat, and that it is bounded by deep ocean. So the size of a steady-state ice sheet must be equal to the size of the continent! The geometry is sketched in Figure 8.1.

125

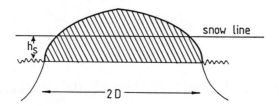

Figure 8.1. A perfectly plastic ice sheet on a bounded continent.

We first note that if $h_s > 0$, a continent without an ice sheet represents a stable steady state. Whether an ice sheet can be in a stable equilibrium depends on its average mass balance, of course. If the mass balance increases linearly with height, the total mass balance of the ice sheet is proportional to the mean surface elevation $\bar{H}$ (note that we consider a flat continent and ignore bedrock sinking). A stable ice sheet is thus possible only when $\bar{H} > h_s$. This observation leads to the solution diagram of Figure 8.2. Stable steady states are indicated by heavily drawn lines. Two critical points appear: at $h_s = 0$ and at $h_s = \bar{H}$. An expression for $\bar{H}$ is easily found by integrating (3.4.4):

$$(8.1.1) \qquad \bar{H} = \frac{1}{D} \int_0^D \sigma \sqrt{x} \, dx = \frac{2}{3} \sigma \sqrt{D} \quad .$$

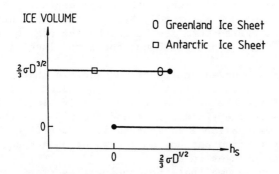

Figure 8.2. Stable steady states for a perfectly plastic ice sheet on a bounded continent. Climatic conditions are represented by the snow-line elevation h_s.

This simplified picture of a continental ice sheet shows how
boundary conditions may lead to a strongly nonlinear response
to changes in environmental conditions, in this case represented
by changes in h_S. The fact that the mass balance depends on
surface elevation is very important, because it enables an ice
sheet to survive a period of increased snow-line elevation. The
Greenland Ice Sheet is a good example: the average snow-line
elevation is about 1500 m, implying that if under the present
climatic conditions the ice would be removed, the ice sheet would
not recover. In contrast, this does not apply to the (main part
of the) Antarctic Ice Sheet. Conditions are much colder there,
and an ice sheet would certainly form again. Although the
analysis carried out in this section is simple and schematic,
it nevertheless enables us to see clearly the difference between
the present two major ice sheets, as indicated in Figure 8.2.

8.2 A semi-infinite continent and a sloping snow line

Next we consider a situation in which a continent is bounded
at one side only. It is easy to see that now an ice sheet of
finite size is not possible when h_S is constant: there will be
no ice sheet or one that grows to infinity. A steady-state ice
sheet becomes possible, however, when the height of the snow line
increases with distance to the edge of the continent. We then
have a situation typical for the American and Eurasian ice sheets
as they appear during ice ages, see Figure 8.3. To find the
equilibria for this configuration, we once more use the perfectly
plastic model.

First we specify the mass balance as

$$(8.2.1) \qquad M = a(x-P) + bz \quad ,$$

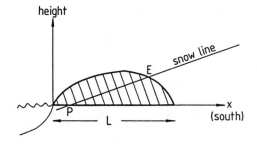

Figure 8.3. A typical northern hemisphere ice sheet.
E is the equilibrium point, P the climate point.

where a, b and P are constants. We take $a < 0$, so M increases
linearly with both surface elevation z and x (distance to the
edge of the continent). The snow line is given by M=0, so

(8.2.2) $h_s = - a(x-P)/b$.

It follows that $-a/b$ is the slope of the snow line, and that at
x=P the snow line intersects sea level. We call this point P the
climate point, because variations in climatic conditions can in
a first approximation be represented by shifting P along the
x-axis. When a and b are constants, this is equivalent to moving
the snow line up and down.
   The total mass balance of the ice sheet is determined mainly
by the location of the equilibrium point E, defined as the
intersection of snow line and ice-sheet surface. This again
demonstrates the importance of the coupling between mass balance
and surface elevation: when P is constant, i.e. when climatic
conditions do not change, the position of the equilibrium point
may still vary significantly.
   For the ice sheet depicted in Figure 8.3, steady-state
conditions imply that the average mass balance of the southern
half of the ice sheet, denoted by M^*, is zero. The rate of
accumulation over the northern half equals the production of
icebergs at x=0. Hence the steady states are found by setting
$M^*(L)=0$. When $dM^*/dL < 0$ the steady state is stable, and when
$dM^*/dL > 0$ it is unstable.

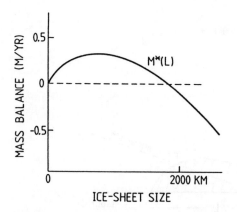

Figure 8.4. Relation between mean mass balance and
ice-sheet size when the climate point is located at
the northern edge of the continent.

$M^*(L)$ is obtained by integrating $M(x,z)$ along the surface of the ice sheet. So

$$(8.2.3) \qquad M^*(L) = \frac{2}{L} \int_{\frac{1}{2}L}^{L} \{a(x-P) + b\sigma(L-x)^{\frac{1}{2}}\} \, dx =$$

$$= B_1 + B_2 L^{\frac{1}{2}} + B_3 L \quad ,$$

$$\text{where} \quad B_1 = -aP \quad ,$$

$$B_2 = 0.47 \, b \quad ,$$

$$B_3 = 0.75 \, a \quad .$$

It is instructive to have a closer look at $M^*(L)$. We first note that, if the mass balance decreases in southward direction ($a < 0$), M^* always becomes negative for sufficiently large L. So the ice sheet is always bounded. In Figure 8.4 a typical M^* versus L curve is shown for the case that $P=0$. This is the situation in which the climate point is located at the northern shore of the continent. According to (8.2.3), changes in P just shift this curve up and down.

If P increases the curve moves downward and two equilibrium states appear, the smaller one being unstable. It is obvious that a critical value of P exists, P_{cr}, at which these equilibria disappear. For negative values of P, the $M^*(L)$-curve shifts upward, and only one stable equilibrium is possible. In addition to this, $L=0$ always represents a stable equilibrium whenever $P < 0$.

The equilibrium values of L are found by solving (8.2.3), yielding

$$(8.2.4) \qquad L = \frac{B_1^2 - 2B_1 B_3 \pm \sqrt{B_2^2 - 4B_1 B_3}}{2B_3^2} \quad .$$

A solution diagram showing L as a function of the position of the climate point is given in Figure 8.5. Curves are drawn for various values of the slope of the snow line. Apart from $L=0$, the solution diagram is a parabola which is tangent to the line $L=0$ at $P=0$. The critical points (black spots in the figure) are given by

$$(8.2.5) \qquad P_{cr} = -0.074 \, (\sigma b/a)^2 \quad .$$

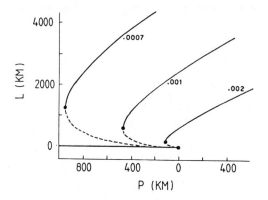

Figure 8.5. Solution diagram for a northern hemisphere
ice sheet. Solid lines represent stable equilibria,
broken lines unstable equilibria. Labels on the curves
give the slope of the snow line. Other parameter
values: b=0.001 yr^{-1}, σ=2.5 m$^{\frac{1}{2}}$.

So the parameters determining the 'magnitude' of the hysteresis
are the yield stress and the slope of the snow line (-a/b).
  The picture emerging here is rather similar to that of
Figure 8.2: for very cold conditions there must be an ice sheet,
for very warm conditions a steady-state ice sheet is not possible,
and in between both a large ice sheet and no ice sheet represent
equilibrium situations.
  The value of σ used in the computation of the solution diagram
is rather small. Still a very large ice sheet appears for
moderately cold conditions. The reason for this can be found in
the fact that the mass balance increases linearly with height.
In reality snow accumulation is strongly limited by low air
temperatures, and in the schematic treatment given above this
point has not been taken into account. Later on, in more specific
modelling of the northern hemisphere ice sheets, snow accumulation
will be subject to an upper limit. This does not change the
qualitative behaviour of the ice sheets, however. Another approach
is to use constant rates of accumulation above the snow line and
ablation below the snow line (Weertman, 1961). This yields a
similar form of M^{x}(L).
  Although a discussion on ice-age theories will follow in a
later chapter, we can already conclude from Figure 8.5 that
the response of the northern hemisphere ice sheets to a varying
snow-line elevation may be very complicated. This is particularly
true when the characteristic growth time of the ice sheets is
comparable to the time scale on which the snow line moves up and
down. With regard to changes in snow-line height associated with

the Milankovitch insolation variations this appears to be the
case. So any ice-age theory in which ice volume, or rate of growth
of ice volume, is set proportional to insolation should be
expected to fail.

Finally we should mention that the importance of the feedback
between surface elevation and mass balance for a growing
continental ice sheet was already stressed by Bodvarsson (1955).
Weertman (1961) first demonstrated in a quantitative way the
resulting hysteresis for a northern hemisphere ice sheet.

8.3 Linear perturbations

In this chapter, we so far only considered steady states. Far
from critical points, the response of an ice sheet to changes in
external conditions can be investigated by linearizing about a
proper reference state. As an example we consider an ice sheet
on a bounded continent, as in section 8.1. Instead of using the
perfectly plastic model, we now employ a simple flow law.
    In order to obtain a 'zero-dimensional' ice-sheet model, we
will insert typical quantities in the appropriate equations.
Another possibility would be to start with a steady-state profile
as shown in Figure 3.5, and linearize around such a state.
However, we like to relate such bulk quantities as mean mass
balance and mean ice thickness, and in that case simple scaling
should work.
    We consider an ice sheet of half width (or radius) L, typical
ice thickness H_0, and mean mass balance M_0. Inserting these
quantities in the continuity equation (3.5.2) we obtain

$$(8.3.1) \qquad \frac{dH_0}{dt} = - \frac{H_0 V_0}{L} + M_0 \quad .$$

V_0 is a typical ice velocity. To express V_0 in H_0 and L we use
(3.5.3) and (3.5.5), yielding

$$(8.3.2) \qquad V_0 = C \, (H_0^2/L)^m \quad ,$$

where C is a generalized flow constant. Combining these
equations then gives the rate of change of typical ice thickness:

$$(8.3.3) \qquad \frac{dH_0}{dt} = - C \, \frac{H_0^{2m+1}}{L^{m+1}} + M_0 \quad .$$

C is generally chosen in such a way that, for given climatic
conditions and ice-sheet size, a reasonable typical ice thickness
is obtained. Ice thickness as a function of the mean mass balance
is found from (8.3.3) by setting $dH_0/dt=0$. So, for $m=3$,

$$(8.3.4) \qquad H_o = C^{-1/7} \, M_o^{1/7} \, L^{4/7} \ .$$

Figure 8.6 shows a plot of $H_o(M_o)$ for $C=1$ and $C=0.1$ m^{-2} yr^{-1}.
Except for low values of the mass balance, the dependence of
H_o on M_o is not very strong (recall that for a perfectly plastic
ice sheet H_o is independent of the mass balance).
   Next we consider an ice sheet that is in a steady state
according to (8.3.4). If M_o changes, the ice sheet will react in
order to achieve a new balance between ice-mass discharge and
accumulation. In case of fixed L, the only way to restore the
balance is a change in H_o. The response of the ice sheet can thus
be obtained from linearization around the steady state. Writing
$H'=H-H_o$ and $M'=M-M_o$, where primes indicate perturbations, the
linearized equation becomes

$$(8.3.5) \qquad \frac{dH'}{dt} = -(2m+1) \, H_o^{2m} \, L^{-(m+1)} \, H' = -\frac{H'}{T_r} + M' \ .$$

The relaxation time T_r equals

$$(8.3.6) \qquad T_r = \frac{1}{C \, (2m+1)} \, L^{m+1} \, H_o^{-2m} \ .$$

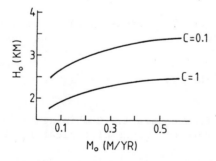

Figure 8.6. Equilibrium ice thickness as a function
of the mass balance, for two values of the generalized
flow constant C (in m^{-2} yr^{-1}).

From this analysis it thus appears that the relaxation time increases with ice-sheet size and decreases with ice thickness. But H_0 and L are coupled, of course. For L=1000 km, H_0=2500 m, m=3 and C=0.1 m^{-2} yr^{-1}, we find T_r=5851 yr, which can be considered as a typical value.

It is important to realize that T_r is a relaxation time, not a characteristic time for ice sheet growth or decay. Equation (8.3.6) can only be used to estimate the time an ice sheet needs to react to a small change in environmental conditions. Small here means that the change in environmental conditions does not bring the corresponding steady state close to a critical point.

Another remark concerns the assumption of constant ice-sheet size. In many situations an ice sheet, or a drainage basin of an ice sheet, may react by a change in ice thickness as well as by movement of the ice-sheet edge. Such a situation is more difficult to analyze by simple methods.

A possible futher application of (8.3.5) concerns the reaction of an ice sheet to random forcing. The climate system contains a considerable amount of internal variability and consequently the mass balance of an ice sheet will vary in an, at least seemingly, irregular way. Generally the time scale of such variations in M will be much smaller than the relaxation time of the ice sheet being forced. So we expect that the ice sheet integrates the fluctuations in mass balance.

An example of the behaviour of a stochastically forced ice sheet is shown in Figure 8.7. The picture was obtained by integrating (8.3.5) in time with M'(t) being a random process (white noise), simulated with the aid of a random number generator T_r was set to 5000 yr. In the figure the effect of the large

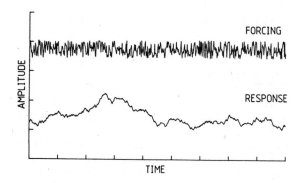

Figure 8.7. Response of an ice sheet (lower curve) to random fluctuations in the mass balance (upper curve). Scales are arbitrary.

response time of the ice sheet is clearly seen. The bulk of the
variance of ice thickness is contained in the longer time scales.

A more general formulation of the (linear) response of an ice
sheet to variations in M is obtained by means of a Fourier
transform of (8.3.5). Denoting a Fourier-transformed variable by
a tilde, we find (ω is frequency)

$$(8.3.7) \qquad i\omega\tilde{H}'(\omega) = -\frac{\tilde{H}'(\omega)}{T_r} + \tilde{M}'(\omega) \quad ,$$

from which it follows that the power spectrum of H' is

$$(8.3.8) \qquad S_{H'}(\omega) = \frac{|M'(\omega)|^2}{\omega^2 + 1/T_r^2} \quad .$$

If M' is a white-noise process its spectral level $|M'|^2$ is constan
constant and S_H reduces to red noise, i.e. the spectrum drops off
with frequency as $1/\omega^2$. The total variance of H', or its standard
deviation σ_H, is obtained by integrating S_H over frequency. So

$$(8.3.9) \qquad \sigma_H^2 = \frac{1}{\pi} t^* |M'|^2 \int_\infty^\infty (\omega^2 + 1/T_r^2) \, d\omega = t^* T_r \sigma_M^2 \quad .$$

Here we have used $\sigma_M^2 = t^*|M'|^2$, where t^* is an averaging time
for the mass balance M'.

From (8.3.9) we see that the total variance of the ice
thickness increases with relaxation time. A simple calculation
shows that a standard deviation of 1 cm ice depth / yr in the
mass balance of an ice sheet with a relaxation time of 5000 yr,
leads to a 0.5 m standard deviation in ice thickness.

8.4   The effect of orography

Since the annual ice-accumulation rate increase strongly with
elevation, it is obvious that the presence of high grounds will
favour the initiation of an ice sheet. In the same way as a
continental edge, where H is precribed to be zero, leads to
multiple steady states (Figures 8.2 and 8.5), the presence of a
mountain range may lead to bifurcation of the steady state
solution. In that case the role of the infinite sink for ice at
the edge of the continent is taken over by the very rapidly
increasing ablation when the ice flows down the mountain slopes.

To illustrate the effect of a mountain range we consider a few

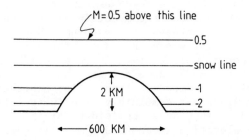

Figure 8.8. Mass balance field (m ice depth / yr) and topography used to study the effect of a mountain range.

experiments with the numerical ice-sheet model discussed in sections 4.1-4.2. Figure 8.8 shows the geometry for this experiment. A mountain range 600 km wide and 2 km high is placed in the centre of the model domain. The lines of equal mass balance are horizontal according to $M=c_1(z-Z_s) + c_2(z-Z_s)^2$, where c_1 and c_2 are constants, and Z_s is the elevation of the snow line. So dM/dz decreases when going upward, and becomes constant at a value of 0.5 m/yr.

The aim of the experiment now is to find out for what values of Z_s ice sheets in equilibrium are possible. Starting from various initial conditions the numerical ice-sheet model produces stable steady-states as summarized in Figure 8.9. The ice-sheet size is given in terms of 'two-dimensional volume', being the product of mean ice thickness and length of the sheet.

As expected, a stable ice cap is possible even when the snow line is above the summit of the mountain range. For $Z_s>2300$ m,

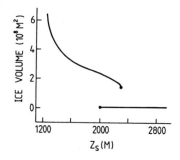

Figure 8.9. Stable equilibrium states for the mountain range experiment. The summit of the mountain has an elevation of 2000 m.

however, the height - mass balance feedback cannot keep pace with the large ice-mass discharge down the mountain slopes and the ice cap disappears. The strong increase of ice volume when Z_s drops below 1600 m markes the point where the edges of the ice cap reach the flat land. For snow-line elevations below 1300 m the ice sheet grows out of the model domain (note that in case of a snow line with zero slope, an ice sheet on a flat unbounded continent will either disappear completely or grow to infinity).

In the experiment described here the 'width' of the hysteresis in terms of snow-line elevation is about 300 m. This value depends on the model parameters c_1, c_2, the flow constant, and the shape of the mountain. A larger flow constant and a steeper mountain will tend to narrow the width of the hysteresis. We may thus conclude that one high plateau is more important than a few high but steep mountain ranges.

9. MORE ABOUT ENVIRONMENTAL CONDITIONS

In our modelling efforts the environment of an ice sheet has
been formulated in a very simple way. The ice accumulation rate
has been prescribed as a simple function of elevation with
respect to the snow line, which itself has been assumed to be a
linear function of position. In the first sections of this
chapter we will discuss in more detail the basis of such
approximations, and see that depending on local conditions a
more sophisticated parameterization of the mass balance is
sometimes required.

In the beginning of this book attention was paid to the
ice-albedo feedback on the global scale. Since the models used
were zero-dimensional, the interaction between ice sheets and
the climatic environment could be studied only in a very
schematic way. Including at least latitude dependence gives the
possibility of studying more explicitly this interaction. The
second part of this chapter will discuss results of a latitude
dependent climate model, based on the energy balance (2.2.2), in
which ice sheets can be included.

9.1 Ablation

The annual ice-accumulation rate is the resultant of
precipitation (snow or rain), evaporation, sublimation and
run-off. A large amount of run-off is only possible when melting
rates are high. Except for very cold conditions, prevailing for
example over the central part of the Antarctic Ice Sheet,
accumulation is mainly by precipitation and ablation mainly by
melting.

The high surface temperatures needed to melt snow or ice may
be established in several ways. In many situations melting occurs
at a high rate in short intervals, making it difficult to model.
To calculate annual ablation from annual mean climatic variables
is hardly possible; at least the yearly cycle should be taken
into account.

Apart from freezing of water, the surface gains heat by
incoming solar radiation (which is partly reflected), infrared
radiation originating in the atmosphere and exchange of sensible

137

heat by turbulent transport in the atmospheric boundary layer.
It depends on the local geographical and climatic conditions
which process is decisive in creating surface temperatures above
the melting point. An important factor also is the speed at which
heat is transported downward in the snow or ice layer. In ice
heat is much easier conducted than in snow.

At high latitudes, where the radiation balance of the surface
is generally strongly negative, the atmospheric boundary layer
is characterized by the presence of a strong inversion: the
potential temperature increases with height (see Figure 9.1).
The stable stratification suppresses vertical motions in the
boundary layer, and sensible heat transfer is virtually absent.
When in such conditions insolation during the day is high, the
heat budget of the surface becomes positive, surface temperature
rises and convection will be initiated just above the surface.
In the growing convective layer potential temperature is
virtually constant (implying a temperature decrease of 1 K per
100 m). If the amount of absorbed radiation is large enough, the
surface may reach the melting point and some meltwater may be
produced. By the end of the day however, the surface heat
balance rapidly becomes negative.

The stronger the atmospheric inversion, the larger the
probability that the ice/snow surface will reach the melting
point, as can be concluded from the arguments given above.
Another factor contributing to this is the infrared radiation
from the atmosphere. When warm air is overlying the cold surface
layer, the net energy loss from the surface by infrared radiation
may be substantially reduced. It is nevertheless questionable
whether clear-sky conditions with high insolation contribute
significantly to melting. Due to the radiative properties of
snow (high reflectivity for solar radiation, emission of
radiation at wavelengths for which the atmospheric absorptivity
is comparatively small) it requires large amounts of incident

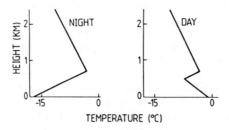

Figure 9.1. Stable boundary layer over an ice sheet,
typical for clear-sky conditions. When during the day
insolation is high, surface temperature increases
strongly because heat is trapped below the inversion
(right part of the figure).

solar radiation to make the surface heat balance positive. Once
the snow is melted away and bare ice (or dirty snow) is at the
surface, a larger fraction of the solar radiation is absorbed
and melting rates may become significant.

Conditions much more efficient in establishing high melting
rates, in particular when snow cover is still present, are
associated with intrusions of warm air masses originating at
lower latitudes. Large amounts of warm air may be swept poleward
by developing baroclinic waves (see Figure 9.2). The accompanying
high wind speeds will erode the temperature inversion and a
large downward flux of sensible heat becomes possible. Since such
air masses are generally rising, extensive cloud cover is present,
radiating large amounts of energy downwards. In this type of
situation the radiation balance of the surface can easily become
positive, even during the night. So warm air invasions appear as
important events with regard to ablation at the surface of an
ice sheet.

The actual loss of ice by melting is in fact determined by
the run-off. Since in many situations melt water is produced
during warm periods of short duration, the speed at which the
water can be removed from the ice sheet is the crucial factor.
In view of this, and the fact that surface temperature decreases
strongly with elevation, it is not surprising that the bulk of
ablation takes place on the edges of an ice sheet.

Although calculations of snow and ice melt have been carried
out on the basis of sophisticated models of the surface energy
balance (sometimes coupled to a model of the atmospheric
boundary layer), such procedures do not appear very suitable for
paleoclimatic studies. One reason is that they require large
amounts of computer time, another one that detailed models
require detailed boundary conditions which are in most cases not
available.

An alternative way to calculate ice/snow melt is to employ

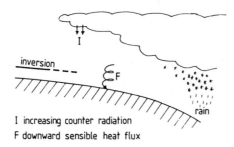

I increasing counter radiation
F downward sensible heat flux

Figure 9.2. Illustration of processes that may lead to
melting of snow and ice during an invasion of warm air
masses.

observational data. Many measurements on glaciers and small ice caps have been carried out, and recently Pollard (1980) attempted to derive from such measurements a relationship between monthly snow/ice melt and monthly surface air temperature T_m and insolation Q_m. His expression reads

(9.1.1)   Melt = max $(0; 0.1 T_m + 0.003 Q_m - 0.5)$.

Here T_m is in $^\circ C$, Q_m in W/m and the amount of melt in m/month (water). When it is desirable to take into account the effect of varying surface albedo, the insolation term can be replaced by the radiation balance at the surface, with an appropriate change in the constant of (9.1.1).

Although the parameterization of the ablation rate as given above is probably the best general relationship one can obtain from observational data, it should be employed carefully. There is not much evidence that present conditions on glaciers are representative for conditions prevailing in the ablation zones of large ice sheets.

9.2   Accumulation

Accumulation is not easier to model than ablation. First of all air temperature is a very important factor. Air temperature at the surface determines whether precipitation falls in the form of snow or rain. Also, the amount of water vapour that can be

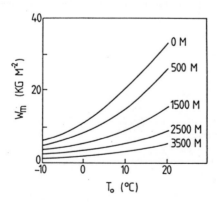

Figure 9.3. The amount of precipitable water in the atmosphere as a function of sea-level temperature, for various surface elevations.

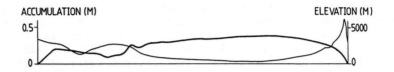

Figure 9.4. A cross section through the Antarctic Ice
Sheet, showing the relation between surface slope and
annual snow accumulation. From Atlas Antarktiki (1966).

contained in the atmosphere strongly depends on temperature. Over
the present Antarctic Ice Sheet for example, the very low air
temperature does not permit a larger precipitation rate than
0.17 m/yr (mean value over the entire ice sheet). This point is
further illustrated in Figure 9.3, where the amount of maximum
precipitable water in the atmosphere is given as a function of
surface temperature. In calculating this figure standard profiles
of air temperature were used. Results are shown for various
surface elevations. In view of this picture is not surprising
that in the central part of the Antarctic Ice Sheet accumulation
drops below 0.05 m/yr, while in south central Greenland a value
of about 0.4 m/yr is found.
     Another important factor is enhancement of accumulation by
orographic effects. Mountain ranges generally experience
precipitation amounts much higher than the surroundings, which
is due to forced uplift of moist air. We should thus expect that
near the edge of an ice sheet, where the surface slope is large,
precipitation rates will reach a maximum value. Observations
confirm this, as is illustrated in Figure 9.4. This figure shows
a cross section of the Antarctic Ice Sheet and gives surface
elevation and annual accumulation. It is quite obvious that
accumulation and surface slope are strongly coupled.
     When an ice sheet shrinks or expands, the zone of high
precipitation rates will shift with the ice-sheet edge. This
mechanism will favour growth of the ice sheet in the direction
of the mean wind. Below we discuss a model experiment on the
Scandinavian Ice Sheet, in which the effect of orographically
induced precipitation is calculated explicitly (Sanberg and
Oerlemans, 1983).
     A calculation of precipitation should preferably be based on
a balance equation for precipitable water, for example

$$(9.2.1) \qquad \frac{\partial W}{\partial t} = -\vec{v}.\nabla W + (W_m-W)/T^* - (f_0+f_1 S)\dot{W} + D\nabla^2 W .$$

The first term on the right-hand side represents advection of
precipitable water by the horizontal wind v (in the lower
troposphere). The second term describes a source: evaporation
from the surface. The atmosphere tends to become saturated
($W=W_m$) on a time scale T^* (typically 3 days over sea, 6 days over
land, very large over an ice or snow surface). The third term
represents precipitation and consists of a constant part f_0W,
where f_0 is determined by the general climatic conditions, and
an orographic part f_1SW, where S is the component of the surface
slope in the wind direction. S should also satisfy the condition
S > 0, so precipitation at the leeward side equals the background
value f_0W. The last term in (9.2.1) takes care of horizontal
mixing of precipitable water.

Eq. (9.2.1) can be solved numerically when the wind field $\vec{v}$,
surface temperature T_s (to calculate W_m), the constants f_0, f_1
and D, and the topography (needed to calculate W_m and S) are
known. When boundary conditions are constant in time the
solution will converge to a steady state. To determine values
of the constants mentioned above one can try to find the best
simulation of a present-day annual precipitation pattern.
Figure 9.5 shows how this approach works for Europe. The
precipitation model (9.2.1) was solved on a 100x100 km grid with
prescribed present-day topography, annual surface temperature and
a westerly wind of 10 m/s. Although discrepancies between
observations and model results exist without doubt, the pattern
of high precipitation rates in mountain ranges is simulated well.
The general decrease of precipitation towards the northeast is
also present in the model result. Errors are significant in

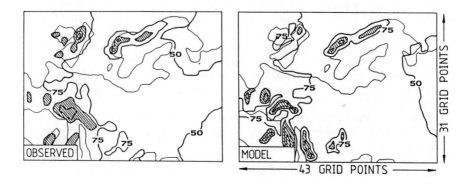

Figure 9.5. A comparison of annual precipitation over
Europe, as observed (left panel), and calculated with
(9.2.1). The contour interval is 0.25 m. Amounts over
1 m are shaded. From Sanberg and Oerlemans (1983).

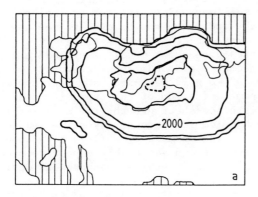

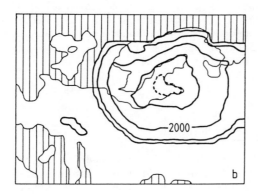

Figure 9.6. Simulation of the growth of the Scandinavian
Ice Sheet. The pictures show ice thickness after
45 000 yr of simulated time (the contour interval is
1000 m). In case b, the effect of the shape of the
ice sheet on the precipitation pattern has been
'switched off'. In both experiments, climatic conditions
were changed by a 7.5 K drop in annual mean temperature
and a 25 % reduction of the annual temperature range.
The (prescribed) mean wind was 10 m/s from the west.
From Sanberg and Oerlemans (1983).

southern Europe, probably because here the downward branch of
the subtropical high-pressure cell will tend to create dry
conditions – an effect not taken into account.

Having derived in this way values for the precipitation
factors f_0 and f_1 , the model can now be used to see how
orographic effects modify the growth of an ice sheet.
To this end the numerical ice-sheet model described in section
4.1 was run for the geometrical conditions presently found in
Europe. The same 100x100 km grid was used, and ice/snow melt was
calculated from a parameterization of the type discussed in
section 9.1. To initiate ice-sheet growth, the annual temperature
at 70 ^{O}N was set 7.5 K below its present-day value and the
annual temperature range was reduced by 25%. This is about the
largest climatic cooling one can reasonably expect to occur.
For more details on this experiment, see Sanberg and Oerlemans
(1983).

Figure 9.6a shows the steady-state distribution of ice
thickness, reached after 45 000 years. The shape of the ice
sheet then resembles that known from the Riss ice age (e.g. Lamb,
1977). Figure 9.6b corresponds to a run in which the effect of
the varying topography, associated with the evolution of the
ice sheet, was not taken into account in the calculation of the
accumulation rate. The comparison between Figures 9.6a and b
clearly shows the effect of upslope precipitation: it causes the
ice sheet to grow in westerly direction, thus covering the
entire North Sea region.

9.3   The zonal mean energy balance

As we have seen in chapter 2, the large albedo of ice and snow
considerably enhances the response of the climate system to
perturbations of the energy budget. A more quantitative
assessment of the temperature – ice – albedo feedback requires
a consideration of at least the latitude-dependent (i.e. zonal
mean) climatic state. This implies that in the energy balance
meridional heat fluxes have to be taken into account.

A class of models widely used to study the albedo feedback is
based on the energy balance

$$(9.3.1) \qquad Q(\phi)\ \{1-\alpha(\phi)\} + D\nabla^2 T(\phi) = I(\phi) \quad .$$

Here all quantities are zonal averages depending on the
geographical latitude ϕ. The first term represents the absorbed
solar radiation, the second one divergence of the dynamical heat
fluxes, and $I(\phi)$ the net outgoing infrared radiation. The heat
flux by atmospheric motions and ocean currents is set

proportional to the meridional temperature gradient, leading to
a diffusion-type term in the energy balance. Absorbing a factor
one over the earth's radius squared in the energy diffusivity D,
the Laplace operator for this problem is

(9.3.2) $\nabla^2 = \partial^2/\partial\phi^2 - \tan(\phi)\ \partial/\partial\phi$.

Expression (9.3.2) can easily be derived by integrating the
Laplace operator in spherical coordinates over longitude.
    The outgoing infrared radiation I can be parameterized as

(9.3.3) $I = a + b(T-\gamma h) = a + bT - b\gamma h$.

Here a and b are constants (see the discussion in sections 1.1
and 2.2) and T-γh is the annual surface temperature, calculated
from the sea-level temperature T by using a lapse rate γ (the
zonal mean surface elevation is denoted by h). Substituting
(9.3.3) in (9.3.1) yields

(9.3.4) $(\nabla^2-b)T = -Q(1-\alpha) + a - b\gamma h$.

When insolation, planetary albedo and surface elevation are known,
(9.3.4) can be solved for T(ϕ).
    One way to do this is to write (9.3.4) for a grid, with grid
points for example at $\phi_i = -87.5 + 5(i-1)$, where i runs from 1 to
36. Negative latitudes then refer to the southern hemisphere.
So (9.3.4) is transformed into a set of 36 coupled linear
equations in T(ϕ_i). In compact form:

(9.3.5) $A\ \vec{T} = \vec{f}$,

where $\vec{T}$ is the 'temperature vector' $T(\phi_1),\ldots,T(\phi_{36})$, $\vec{f}$ the
discrete form of the right-hand side of (9.3.4), and A a matrix.
With

$$\mu_i = 1/d^2 - \tan(\phi_i)\ /2d \quad,\ \text{and}\quad \nu = -2/d^2 \quad,$$

where d is the spacing of grid points (5 $^{\circ}$), this matrix takes
the form

$$
A = \begin{bmatrix}
-\mu_1-b & \mu_1 & & & & & \\
\mu_{35} & \nu-b & \mu_2 & & & \emptyset & \\
 & \mu_{34} & \diagdown & \diagdown & & & \\
 & & \diagdown & \diagdown & \diagdown & & \\
 & & & \diagdown & \diagdown & \diagdown & \\
 & & & & \diagdown & \diagdown & \mu_{34} \\
 & \emptyset & & & \mu_2 & \nu-b & \mu_{35} \\
 & & & & & \mu_1 & -\mu_1-b
\end{bmatrix} \quad .
$$

In deriving this matrix the boundary condition $\partial T/\partial \phi$ has been set to zero at the poles. This guarentees that there is no flux of energy over the poles, a condition to be fulfilled in zonal mean models. A is of tridiagonal form and can easily be inverted by standard methods. The solution is then directly obtained from

$$(9.3.7) \qquad \vec{T} = A^{-1} \vec{f} \quad .$$

A result of a calculation with this model is shown in Figure 9.7. Insolation at the top of the atmosphere, planetary albedo and surface elevation are shown in the top part of the figure, and the resulting solution for T in the lower part. The observed zonal mean temperature is also indicated. Apparently, the solution is quite realistic except in the polar regions where temperatures are overestimated by the model. This is partly due to the use of a constant value for D. Included in D is the heat flux accomplished by ocean currents. Since in the polar regions oceanic heat transports are absent or very small, the constant-D formulation overestimates the poleward heat flux at high latitudes and thus leads to polar temperatures that are too high.

Another reason for the deviations may be the pronounced stratification of the atmosphere at high latitudes. The energy-balance model described here is essentially based on the idea that sea-level temperature is somehow representative for the entire atmosphere-ocean column. This assumption is likely to fail when the atmosphere is very stable and consequently vertical exchange is suppressed.

A further comment involves the choice of model parameters used in the calculation of $T(\phi)$. The result was in fact optimized by chosing best values for D (which determines the meridional temperature gradient) and the constant a in the expression for

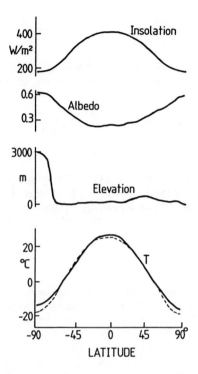

Figure 9.7. Annual sea-level temperature calculated
from the zonal energy balance (lower curve). Observed
temperatures are indicated by the dashed line. The
input to the model is shown in the upper curves.

the infrared emission (determining to a large extent the global
mean temperature). These parameters can easily be varied within
their range of uncertainty. The results shown here were obtained
with $D = 0.62$ W/$(m^2$ K) and $a = 233$ W/m^2.

9.4   Albedo feedback as a latitude-dependent problem

To study the role of ice and snow cover in enhancing climate
sensitivity, we now have to include the albedo feedback, i.e.
α has to be made a function of T. In that case f depends on
temperature and the set of equations becomes nonlinear. However,
since the planetary albedo is bounded, f is also bounded and
(9.3.5) can be solved by the iterative scheme (n denotes the
number of iteration):

(9.4.1) $\vec{T}_{n+1} = A^{-1} \vec{f}(\vec{T}_n)$,

This scheme always converges to a stable equilibrium state. In practice the procedure is stopped when a condition of the type

$$|\vec{T}_{n+1} - \vec{T}_n| < \varepsilon$$

is fulfilled.

To parameterize the albedo in terms of surface temperature, we first split the planetary albedo in contributions from cloud-covered and cloud-free regions (see section 2.2):

(9.4.2) $\alpha = \alpha_{cl} + \alpha_{cs}(1-N)$.

N is cloudiness, α_{cl} cloud albedo and α_{cs} clear-sky albedo. The albedo feedback operates through the clear-sky albedo, of course. To relate α_{cs} to temperature we need to know how ice and snow cover vary with temperature, and also how ice and snow cover modify α_{cs}. Again observational data may help.

A typical result, derived from climatic data on temperature and snow and ice cover (for the moment we only consider sea ice), is shown in Figure 9.8. The fraction of the ocean covered by sea ice, and the fraction of land covered by snow are displayed as a function of surface temperature. There are no jumps in the curves, but rather smooth transitions. This is due to both zonal

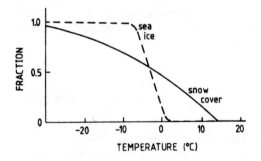

Figure 9.8. Dependence of zonal mean sea ice and snow cover on annual surface temperature, as used in the energy-balance climate model.

asymmetry and the presence of the annual cycle. The latter in
particular makes it possible that for an annual temperature of
10 °C some snow cover still appears. It should be noted
that here 'fraction of a latitude circle covered by snow' and
'fraction of the year with a snow deck' are loosely used as
equivalent concepts, which can be critisized.

Assigning clear-sky albedos to regions covered by snow or ice,
and specifying the fraction of land along a latitude circle, the
contribution of the cloud-free part to the planetary albedo can
now be calculated from T. In climate sensitivity experiments,
one can either keep cloud fixed or make it a prescribed function
of temperature. In the few experiments discussed below the
contribution from clouds is prescribed to be the present one, and
thus is a fixed function of latitude. With clear-sky albedos of
0.22 over land, 0.13 over sea, 0.56 over ice cover, and 0.62 over
snow cover, the planetary albedo is well simulated, see Figure
9.9a.

Having constructed a scheme in which temperature controls the
clear-sky albedo, we are now able to carry out sensitivity
experiments with ice/snow - albedo feedback included. One way
to do this is to vary the solar constant, i.e. to change $Q(\phi)$
with the same fraction at every latitude. A 2% drop of the solar
constant, for instance, leads to a decrease in sea-level
temperature as shown in Figure 9.9b. The drop in global mean
temperature is about 2.4 K. Although the albedo feedback varies
strongly with latitude (it will be strongest near the present-
day snow and ice boundary), the response to changing insolation
is remarkably uniform. This shows that redistribution of energy
in the climate model is very effective, and that the model
essentially behaves in a 'zero-dimensional way'.

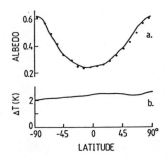

Figure 9.9. Simulation of the present-day planetary
albedo together with observed values (dots), and the
temperature decrease produced by the energy-balance
model when insolation is reduced by 2 %.

A further reduction of the solar constant, by about 10%, leads to transition to a completely snow-covered earth. Increasing the solar constant with a snow-covered earth as initial condition, on the other hand, shows that insolation should be about 10% higher than today to create 'normal' climatic conditions again. So hysteresis occurs as a consequence of the albedo feedback, but we should realize that it is very unlikely that the climate model described here can be applied to conditions differing so markedly from the present state. It is also important to note that the appearence of two stable steady states for a range of solar constants should not be interpreted as a glacial and interglacial regime. Even during a full glacial the earth is still very far from total ice/snow cover.

9.5 Including continental ice sheets

So far we did not consider continental ice in the energy-balance climate model, only sea ice and snow were included. To incorporate continental ice sheets, we first restrict the model to the northern hemisphere, which is easily accomplished by setting $\partial T/\partial \phi = 0$ at the equator and retaining only 18 grid points in the calculations. So we deal with one northern hemisphere ice sheet only.

It would be rather unrealistic to put the southern edge of the

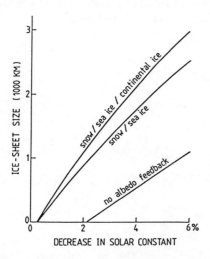

Figure 9.10. Ice-sheet size as a function of the reduction in solar constant, for various forms of the albedo feedback.

ice sheet at some specified temperature. A large continental ice
sheet can penetrate far southward and it seems more reasonable
to specify the temperature of the equilibrium point.
Disregarding for a moment the increase in surface elevation
associated with the ice sheet, and assuming that the mass
balance increases linearly with latitude, the position ϕ_e of the
ice-sheet edge is given by

$$(9.5.1) \qquad \phi_E = \phi_n - \frac{4}{3} \{\phi_n - \phi(T_E)\} \quad .$$

Here ϕ_n is the latitude of the northern shoreline of the
continent ($\simeq 73$ $^\circ$N), and $\phi(T_E)$ the latitude at which the
equilibrium point is found, i.e. at the temperature $T_E (\simeq -10$ $^\circ$C).
It is easily verified that in this case the mean mass balance
of the southern half of the ice sheet is zero. With (9.5.1)
included in the calculation of the clear-sky albedo, the
insolation can again be varied.

An alternative way to display the model response now is to
plot the ice-sheet size, being $(4\phi_E - \phi_n)/3$, as a function of the
imposed decrease of the solar constant. This is done in Figure
9.10. Three cases are shown. First, calculations were made in
which the albedo was kept fixed. In this case a drop in the
solar constant of more than 2% is required to create a (small)
ice sheet. Taking into account the effect of sea ice and snow
cover, corresponding to the experiments described in the
foregoing section, enhances the sensitivity in a substantial way:
a slight decrease in insolation initiates an ice sheet, which
reaches ice-age size for a decrease of 6%. Including the effect
of the continental ice sheet on the albedo appears to have a
moderate effect, the reason being that the surface of the ice
sheet is almost completely covered by snow. Only for larger
drops in the solar constant the southern tip of the ice sheet
penetrates to regions where snow cover is absent in summer, thus
contributing to the increase in annual mean albedo.

Including ice sheets as if they were flat is of course not
very realistic. The next step is to prescribe the ice-sheet
profile, calculate the temperature at the surface of the ice
sheet (using the lapse rate γ) and use this temperature to
determine the position of the equilibrium point ϕ_E. This implies
that the height – mass balance feedback is taken into account.

A useful ice-sheet profile for this purpose is the parabolic
one (L is the size of the ice sheet)

$$(9.5.2) \qquad H(x) = \sigma\{\tfrac{1}{2}L - |x-\tfrac{1}{2}L|\}^{\frac{1}{2}} \quad ,$$

and the temperature along the sheet is thus given by

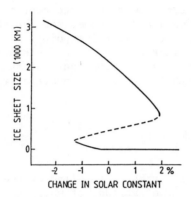

Figure 9.11. Ice-sheet size as a function of the solar
constant. Solid lines represent stable states, the
dashed line unstable states.

(9.5.3) $T(H) = T - \gamma H$.

Repeating the sensitivity experiments with (9.5.2) and (9.5.3)
now leads to a completely different behaviour. Figure 9.11 shows
the result . For a small drop in the solar constant a large ice
sheet appears, which can only be removed by increasing the
insolation to a level 2% above the present one. So again
hysteresis occurs, but in a very different way. Comparing
Figures 9.11 and 8.5 shows a striking resemblence, and it turns
out that the height - mass balance feedback dominates. The
increased sensitivity of the climate model is not caused by the
albedo feedback, but by the effect of the surface elevation of
the ice sheet on the position of the equilibrium point. This very
clearly demonstrates the importance of ice-sheet dynamics in the
global climate system.

A few recent papers on ice sheets in energy-balance climate
models are Pollard (1978), Oerlemans (1980a), Pollard et al.
(1980), Källén et al. (1979), Ghil and Le Treut (1981).

10. THE GLACIAL CYCLES OF THE PLEISTOCENE

The ice ages of the Pleistocene have fascinated quite a number
of scientists from various disciplines, and they still do.
Although it is not the object of this book to detail in depth
with the geological methods that led to the documentation of the
ice.ages as we know it now, we will spend a few pages on how
ideas about the 'glacial theory' evolved, and on what type of
material is most useful in the search for an explanation of the
glacial cycles.

10.1 Observational evidence of ice ages

May be the most appealing signal of the ice ages in the landscape
is the presence of large stones that, according to their apparent
origin, travelled southward over a distance of 1000 km or so. Two
centuries ago the presence of these stones was considered to be
the result of extensive flooding. In the beginning of the
nineteenth century the opinion gradually changed. More and more
traces suggesting former glaciation were found in the Alps and
in Scandinavia. The decisive push towards acceptence of the
glacial theory, as it was called in that time, came from Louis
Agassiz, president of the Swiss Society of Natural Sciences.
In 1837, after a tumultuous debate in a scientific meeting, he
organized an excursion to the Jura and showed that some features
present there could only be the result of extensive glaciation.
Most of the participants, but not all, were convinced. The
definite acceptance of the glacial theory came in 1852, when it
became clear that Greenland was nothing but a huge ice sheet and
it was thus proven that ice sheets with a typical size of 1000 km
are very well possible.
   The wealth of observations gathered by geologists since that
time has provided a rather detailed picture of how far
continental ice sheets in northern America and Europe have
penetrated southwards. Field studies of postglacial uplift and
former sea-level stands have made it possible to estimate the
volume of the American and Eurasian ice sheets during a glacial
maximum (see Table 10.1).
   However, investigation of the landscape on all kinds of signs

of glaciation has one serious disadvantage: the amount of information is distributed very irregular in time. An expanding ice sheet tends to destroy the traces left by its predecessor. Also, dating is not always easy, particularly not before 70 000 years ago (the limit for the C^{14} dating method). The rate of sedimentation varies enormously in time, and sometimes large gaps in the pack of sediments occur. In view of this, the study of ocean sediments that started roughly fifty years ago was very welcome.

A typical value of the rate of sedimentation in the ocean is 1 cm/1000 yr. In the thirties, deep-sea cores were drawn with a typical length of about 1 m, so the analysis of these cores contained information about the climatic history of the last 100 000 years. After the second world war one succeeded in taking cores of more than 10 m long and from these cores we have learned much about the glacial cycles of the last million of years, in particular when it appeared to be possible to find the Brunhes-Matuyama reversal of the earth's magnetic field (about 700 000 years ago) in the cores. Absolute dating of one point was thereby accomplished.

The deep-sea cores can be analyzed in many ways. The abundance of specific planctonic species and oxygen-isotope ratios proved to give very valuable information. In the present context the oxygen-isotope analyses are of great importance, because they measure the total ice volume on earth. This works as follows. When foraminefera living in the ocean water make their shells, they register the H_2O^{18}/H_2O^{16} ratio in the carbonate. The shells sink to the ocean floor and make up a large part of the sediment. So analysis of the O^{18}/O^{16} ratio in deep-sea cores gives an estimate of the relative abundance of H_2O^{18} at the time the carbonate was formed.

The relation between O^{18}/O^{16} ratio and global ice volume stems from the fact that fractionation takes place when water evaporates from the ocean surface (H_2O^{16} evaporates 'easier').

ICE SHEET	VOLUME (10^6 km^3)	
	present	ice age
Antarctic	30.0	34.0
Greenland	2.6	3.5
North American	–	33.0
Eurasian	–	13.3

Table 10.1   Based on Flint (1971) and Drewry (1982).

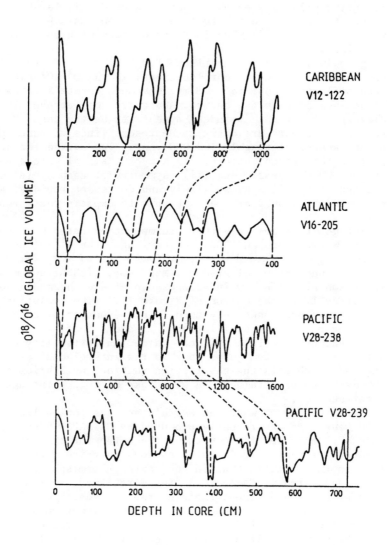

Figure 10.1. Oxygen isotope records from deep-sea cores taken at different locations. The vertical line indicates the last reversal of the earth's magnetic field (not present in V12-122). Dashed lines are lines of equal phase, spaced at roughly 100 000 yr. In the core V28-239 the sedimentation rate is such that the scale gives time before present in thousands of years. The composition is based on Imbrie et al. (1973), and Shackleton and Opdyke (1973, 1976).

The result is obvious: the O^{18}/O^{16}ratio in the ocean will be
comparatively large when the global ice volume is large.
Although there are other effects as well, the main part of
deviations in the O^{18}/O^{16}ratio is associated with variations in
global ice volume.

Figure 10.1 shows a comparison of O^{18}/O^{16}ratios obtained from
deep-sea cores taken at four locations. If the O^{18}/O^{16}ratio
indeed reflects global ice volume, one must expect a large degree
of coherency between the curves. Apart from differences in
sedimentation rate, this indeed appears to be the case. Dashed
lines in the figure are drawn to represent lines of equal phase;
they are spaced at roughly 100 000 years. Two marked features
appear in the curves, namely, (i) a cyclic signal with a period
of about 100 000 years, and (ii) asymmetry of each cycle in the
sense that ice volume grows slowly and decreases rapidly. Any
ice-age theory should at least be able to explain these two
points.

Although it is unlikely that sedimentation rates have been
constant between now and 700 000 years ago, the O^{18}/O^{16}curves
form a continuous record of ice volume being of very large value
for time-dependent model studies. Moreover, without any model
calculation at all a few inferences can already be made when we
look carefully at the curves of Figure 10.1. The evolution of
global ice volume looks very much like the output of a system
exhibiting relaxation oscillations. It seems that, when ice
volume reaches a certain critical value, an instability occurs
and the climate system jumps back to the interglacial state.
Connected to this is the observation that an equilibrium state
is probably never attained. We will take up this point again in
a later section.

A detailed map of what the earth's surface looked like
20 000 years ago was recently produced by CLIMAP (1976). In
Figure 10.2 only sea-ice cover and the extension of continental
ice sheets is shown. It is quite obvious that ice ages are more
marked in the northern than in the southern hemisphere. As was
already suggested in Table 10.1, the North-American (Laurentide)
Ice Sheet was the dominant feature; the ice sheet covering
northern Europe (the Fennoscandian Ice Sheet) was considerably
smaller. The fact that Siberia was essentially free of ice is
most likely the result of very dry conditions prevailing at that
time. In addition to this the planetary atmospheric wave
generated by the Fennoscandian Ice Sheet probably had its southward brache over Siberia (see also section 2.6).

In general ice ages appear more pronounced in the North
Atlantic Ocean region. The limit of pack ice was substantially
displaced to the south and so was the whole North Atlantic
Current sytem. Most workers believe that the moisture supply for
both the Laurentide and Fennoscandian ice sheets came from the
Atlantic Ocean. Again, the comparatively small difference between
present-day and ice-age conditions over the North Pacific Ocean

may be associated with the planetary wave location mentioned
above.

Although it is doubtful that the Laurentide and Fennoscandian
ice sheets have always been in phase, we can assume safely that
the large variations in total ice volume during the Pleistocene
were associated with changes in the extent of these ice sheets.
The Antarctic Ice Sheet also varied in shape, mainly as a
response to changes in sea level, but the corresponding changes
in ice volume were not very dramatic on the global scale (see
also Table 10.1).

The description of ice ages given above is limited, but
nevertheless allows us to define the problem: What mechanism has
been responsible for the regular sequence of glacial cycles that
dominated the earth's climatic history of the last few millions
of years? Many scientists have tried to answer this question,
and the number of proposed theories is very large. They
naturally fall into two classes: theories involving internal
oscillations of the climate system, and theories explaining
glacial cycles as the response of the climate system to external
forcing (changes in incoming energy). Most popular among the
external-forcing theories is the so-called astronomical theory,
in which insolation variations caused by small changes in the
orbit of the earth constitute the driving force. In the following
section the basis of the astronomical theory will be discussed.

With regard to oscillations of the climate system it seems
that only processes involving the ice sheets themselves possess
a time scale large enough to explain the 'observed' glacial

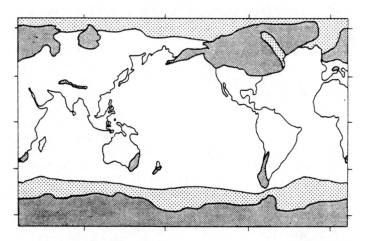

Figure 10.2. Distribution of sea ice (summer, light
shading) and continental ice sheets (heavy shading)
during the last glacial maximum of 20 000 yr ago.
After CLIMAP (1976).

cycles. Even mixing in the ocean, including the deep water, is a rapid process compared to the time ice sheets need to attain ice-age size.

10.2   The astronomical theory

The development of the astronomical ice-age theory started halfway the 19th century. Adhémar was the first one to claim that the occurrence of ice ages was linked to variations in the earth's orbit. Croll, who stressed the importance of the ice-albedo feedback for the first time, and in particular Milankovitch, developed the subject further (the astronomical theory is frequently referred to as the Milankovitch theory). Support for the astronomical theory was recently given by Hays, Imbrie and Shackleton (1976), who showed that power spectra of ice-volume records and insolation variations are rather similar. Before discussing their arguments in more detail, we first turn to the causes of the insolation variations.

The earth is continuously drawn out of its equilibrium path around the sun. Gravitational attraction by the other planets in the solar system causes minor changes in the orbit of the earth. The laws describing this were already known by Keppler. The implications of these laws and the associated insolation variations have been calculated with increasing accuracy, in recent years notably by Brouwer and Van Woerkom (1950), Vernekar (1972) and Berger (1975). The basis of the astronomical insolation variations can be summarized as follows.

Globally speaking, changes in the orbital parameters of the earth of importance with regard to insolation variations are made up of three components, namely,

(i)   precession of the equinoxes. Due to the fact that the earth exhibits precession (the 'point' of the rotation axis moves around in a circle when looking perpendicular to the ecliptic) and the orbit slowly turns within the ecliptic, the seasons effectively move around the orbit. So, depending on the eccentricity of the orbit, the earth – sun distance varies through the year, and the date at which this distance reaches a maximum value changes. The equinoxal precession has an average period of about 22 000 yr, as is illustrated in Figure 10.3.

(ii) variations in eccentricity. The eccentricity of the orbit changes on a very long time scale, a period of roughly 100 000 yr being dominant. The eccentricity varies between 1.00 and 1.05.

(iii) variations in obliquity. The obliquity, defined as the angle between rotation axis and normal to the ecliptic, also varies. Here the major period is about 44 000 yr, and the actual variation is over a range of few degrees.

The present value of the obliquity is 23° 27'.

A complete treatment of how insolation varies as a result of orbital changes would be well beyond the scope of this book. Readers interested in more detail are referred to Berger (1978); here we give a schematic picture only.

As a start we consider the effect of the equinoxal precession. Annual insolation averaged over the earth is not influenced at all by this factor, but a large effect will be present at high latitudes where the bulk of solar energy is received in summer. So at high latitudes, summer and therefore annual insolation is largest when the earth – sun distance is a minimum. This implies that the high latitude insolation variations, as far as equinoxal precession is concerned, are just out of phase in the northern and southern hemisphere. The upper curve in Figure 10.4 shows the effect of equinoxal precession on summer halfyear insolation at 65 °N, say (this is a schematic curve, not based on an actual calculation of insolation!).

We now turn to the variations in eccentricity. It is obvious that equinoxal precession does not affect insolation when the

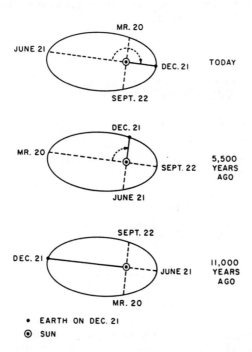

Figure 10.3. Illustration of the equinoxal precession. The seasons travel around the orbit of the earth with a period of about 22 000 yr.
This picture is from Imbrie and Imbrie (1979).

eccentricity is zero - the magnitude of the precession effect is
thus determined by the eccentricity. Combination of the upper
curve in Figure 10.4 with a 100 000 yr periodic signal in the
eccentricity then leads to the second curve (a similar curve was
already derived by Croll). Again this is a schematic curve. It is
worth noting that the eccentricity is the only orbital parameter
causing changes in the annual global insolation. Equinoxal
precession and changes in the obliquity only lead to a
redistribution of solar energy (in space and time).

   Remains to add the effect of changes in obliquity. The
resulting insolation variations are by far the largest at high
latitudes, but now the effect is the same in the northern and
southern hemisphere. An actual calculation for 65 °N, giving
summer halfyear insolation, is shown in Figure 10.4 (lower curve;
Berger, 1978a).

   Much work has been directed towards showing that a direct
relation exists between the occurrence of glacial cycles and
insolation variations. In each study the problem of dating
appears to be crucial. The classical Milankovitch theory
became less popular when C^{14}-dating showed that the waxing and
waning of the ice sheets was a complicated matter and did not
simply follow his insolation curve.

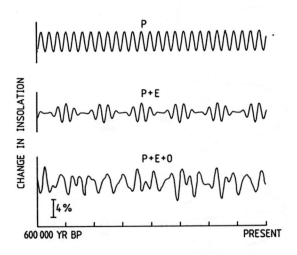

Figure 10.4. The effect of variations in the earth's
orbit on summer insolation at high latitudes. The
upper curve (P) illustrates the effect of precession
(schematic), the middle curve (P+E) shows what happens
when varying eccentricity is added. The lower curve
(P+E+O) shows actual variations at 65 °N for the summer
halfyear (Berger, 1978).

About fifteen years ago, new evidence causing a revival of the astronomical theory came from dating of reef terraces on Barbados, New Guinea and the Hawaiian Islands (Broecker and others (1968)). The formation of these terraces turned out to fit the insolation curve for middle latitudes (where, compared to high latitudes, the obliquity effect is less pronounced).

Hays, Imbrie and Shackleton (1976) presented a statistical analysis of insolation data, ice-volume curves (from δO^{18}records) sea-surface temperature (from radiolarian assemblages in deep-sea cores) and a planctonic species named Cycladophora davisiana, which is abundant when the surface water has low salinity and may thus indicate large-scale melting of ice sheets. After optimising time scales by using various 'age models', power spectra were calculated over the last 480 000 years.

A typical result is shown in Figure 10.5. Spectral peaks occur at frequencies that are roughly in accordance with the periodicities in the insolation variations. This strongly suggests that the occurrence of ice ages and orbital changes are linked in some way. However, the spectral peak near 100 000 years in the insolation variations is rather small, whereas this period is dominating in the ice-volume spectrum. Also, coherency on the 40 000 and 20 000 years time scales is much more convincing than on the 100 000 yr time scale (Kominz and Pisias, 1979). It therefore appears that on the 'shorter' time scales the cryosphere responds linearly to the insolation variations, but that the 100 000 yr power in the climatic variations must be some kind of resonance or nonlinear effect.

Since insolation variations vary with latitude and season, many different insolation curves can be constructed. This introduces many possibilities to fit paleoclimatic data and thus introduces a lot of ambiguity. This does not mean that statistical modelling is useless, but one has to be careful. Straightforward statistical models linking ice-volume records to

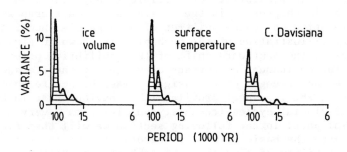

Figure 10.5. Power spectra of ice volume, sea-surface temperature and C. Davisiana. From Hays et al. (1976).

the orbital parameters have been developed by Imbrie and Imbrie
(1980) and Kukla et al. (1981). In the former study it was
shown that models in which the rate of change of ice volume is
set proportional to the perturbation in Northern Hemisphere
summer insolation improve considerably when two time scales are
incorporated: one for ice-sheet growth and one for ice-sheet
decay. They obtained a best fit to the observed ice-volume
record with response times of about 30 000 and 10 000 years
respectively. In view of the essential nonlinearities
(bifurcations) associated with ice sheets in the climate system
(see sections 8.1 and 8.2), this is probably the best one can
get out of a statistical model. A deeper understanding of how
the glacial cycles came about certainly needs modelling of the
essential physics.

Accepting the astronomical theory from a global point of view,
the question to be answered is the following: 'How can the small
insolation variations associated with orbital changes lead to
such dramatic events as the quaternary glacial cycles?'. As
mentioned earlier, many mechanisms have been proposed. The one
that has been studied most is the ice-albedo feedback. In
chapter 2 we have seen that variable ice cover enhances climate
sensitivity and may even create multiple steady states. Latitude-
dependent effects were studied in the previous chapter.

Attempts have been made to refine energy-balance climate
models in such a way that the response to the astronomical
insolation variations attains 'ice-age magnitude'. Such refinements include incorporation of zonal asymmetry (e.g. Oerlemans,
1980) and dealing with the seasonal cycle (e.g. North and
Coakley (1979). Although some authors claim to have constructed
realistic models that exhibit the required sensitivity, the
general feeling is that the ice-albedo feedback is too weak
to explain on its own the glacial cycles.

Another powerful mechanism that enhances climate sensitivity
involves the northern hemisphere ice sheets themselves. The
strong dependence of the mass balance on surface elevation,
leading to the possibility of multiple steady states as
discussed in chapter 8, is the reason why a small drop in snow-
line elevation may cause built-up of a large ice sheet. A
schematic illustration of the mechanism is shown in Figure 10.6.
Apparently, the southward shift of the equilibrium point is due
mainly to the increasing surface elevation of the ice sheet.

Since the strength of the height - mass blance feedback
depends on the ice-sheet profile, it is obvious that in the
study of glacial cycles ice-sheet dynamics form a very
essential part. In the following section we will therefore
investigate the basic effect of bedrock sinking and heat budget
on the stability of a northern hemisphere ice sheet.

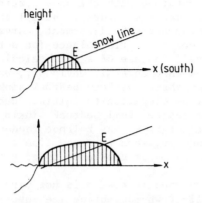

Figure 10.6. The height – mass balance feedback of a
northern hemisphere ice sheet.

10.3 Stability of northern hemisphere ice sheets

As discussed in section 8.2, a comparatively small drop of
the snow line elevation may lead to a large ice sheet. It was
also demonstrated that a very large increase in snow–line
elevation is required to remove a large ice sheet, once it has
been built up. However, bedrock sinking and gradual increase of
the large-scale flow parameter (due to an increase in ice
temperature and basal melting) can 'prepare' an ice sheet for
decay. To study this point in a quantitative way, we employ the
numerical ice-sheet model developed in sections 4.1 (vertically-
integrated ice flow), 7.2 and 7.3 (to calculate bedrock sinking
if desired) and 4.3 (to deal with movement of the snout in case
of bedrock below sea level). We use the geometry of Figure
10.6 with a grid-point spacing of 25 km. The boundary condition
at x = 0 simply is H(t) = 0 (so here the grounding line is
fixed), and the mass balance is parameterized according to

$$(10.3.1) \qquad M = \min \{ 0.35, \ a(h-h_s)+ bx \} \ m \ \text{ice depth/yr.}$$

Here h is surface elevation and h_s snow-line elevation.
The idea now is to initiate ice-sheet growth by lowering the
snow line for some period of time, and then raise the snow line
gradually to see when the model ice sheet becomes unstable and
disappears. By comparing runs with and without bedrock sinking
some insight into the role played by geodynamics can be obtained.
In Figure 10.7 results are shown for a=0.001 yr^{-1} and
b=0.6x10^{-6} yr^{-1} (so the slope of the snow line is 0.0006).

Three cases are shown, all obtained with the same forcing (elevation of the snow line, see the figure). Curve 1 shows the model result in case of no bedrock adjustment, curve 2 in case of damped return to local isostatic balance (on a time scale of 5000 yr), and curve 3 in case of a full calculation of bedrock sinking as discussed in chapter 7. Obviously, bedrock sinking destabilizes the ice sheet. Without bedrock sinking the required increase in snow-line elevation to initiate decay is about 1000 m larger than in case of full bedrock sinking. Note also that the effect of the height – mass balance feedback shows up clearly: even after 20 000 yr, when the snow line at x = 0 rises above sea level, the ice sheet continues to grow for some time.

The maximum ice volume reached in case 3 is not very large. A value typical for the full-grown Laurentide Ice Sheet is two to three times as large. There are various ways to let the model produce a larger ice sheet. The upper bound to the ice accumulation rate (0.35 m ice depth/yr in the results shown here) can be increased, the duration of the cold period can be made longer, the flow constant in the ice-sheet model can be decreased, and, most efficiently, the slope of the snow line can be decreased.

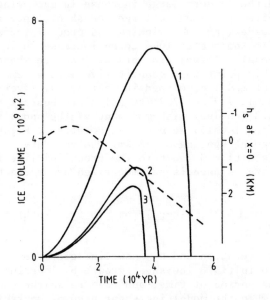

Figure 10.7. Ice volume as a function of time for three cases: (1) no bedrock sinking, (2) local isostatic adjustment, (3) complete calculation of bedrock sinking. Dashed line shows precribed height of the snow line.

In Figure 10.8 results are shown with a snow-line slope of
0.0005. In addition to this modification, the rate of increase
in snow-line elevation after 20 000 yr was halved. Apparently
the maximum ice-sheet size becomes much larger now, and growth
and decay of the ice sheet takes about 65 000 yr. The case
without bedrock sinking is not shown because the ice sheet
became so large that it grew out of the model domain.

   To summarize the effect of bedrock sinking, we can state that
it increases the growth time and decreases the decay time of a
northern hemisphere ice sheet. Decay is particularly rapid when
the rigidity of the lithosphere is taken into account, but even
without this bedrock sinking is a strong destabilizing factor
once an ice sheet is full-grown.

With the experiments of chapter 6 in mind, it is obvious that
the heat budget of an ice sheet also plays an important role.
When an ice sheet forms, conditions are normally cold and the
flow parameter will be comparatively low, thus favouring further
growth. Gradual warming of the ice sheet forms an additional
destabilizing factor, because the increasing deformation rate
and consequent increase in ice-mass discharge will weaken the
height-mass balance feedback. It is not difficult to see that
when basal temperatures reach the melting point the
destabilization can be substantial.

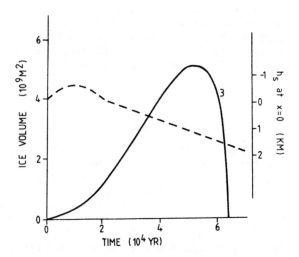

Figure 10.8. Similar to Figure 10.7, but now for a
snow-line slope of 0.0005. Only the run with full
bedrock calculation is shown.

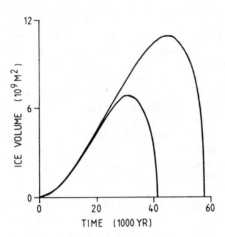

Figure 10.9. Illustration of the effect of an
increasing flow parameter. The upper curve refers to
a run with a constant flow parameter, the lower one
to a run in which the flow parameter increases linearly
with time (a factor 100 in 100 000 yr).

To illustrate this point, two integrations are compared in
Figure 10.9. The first run was carried out with a constant flow
parameter, the second one with a flow parameter increasing
linearly in time. In both cases bedrock sinking was calculated,
and the snow line was prescribed as in Figure 10.8.
The interpretation of the result is simple: an increasing flow
parameter causes the mean ice thickness to decrease, so a larger
part of the surface will become subject to melting.

10.4. Simulation of the ice volume record

Having examined the factors that may destabilize a large ice
sheet, it is of interest to see whether longer model runs with
realistic forcing of the snow line yield glacial cycles that are
similar in character to those actually 'observed'.
   Numerical ice-sheet models with bedrock sinking as only
internal destabilizing factor appear to be able to produce
glacial cycles of long duration (Oerlemans, 1980c; Birchfield
et al., 1981; Pollard, 1982). However, since the effects of
bedrock sinking and gradual warming of the ice sheet are of
roughly equal importance, a proper simulation should deal with
both. In Figure 10.10 some results are shown from long
integrations with a northern hemisphere ice-sheet model that

includes a calculation of temperature - ice flow interaction, and
in which a damped return to local isostatic equilibrium is
assumed. The figure is taken from Oerlemans (1982). Before we
discuss the simulated ice-volume curves, a few remarks on the
model used are in order.

The ice-sheet model is of the type employed in the previous
section, in which a scheme for the calculation of the temperature
field was implemented (the one discussed in section 5.4). Basal
water spreads through advection with the ice velocity (in contrast
to the diffusive treatment given in section 6.4) and affects the
ice-mass discharge. The model is forced by moving the snow line

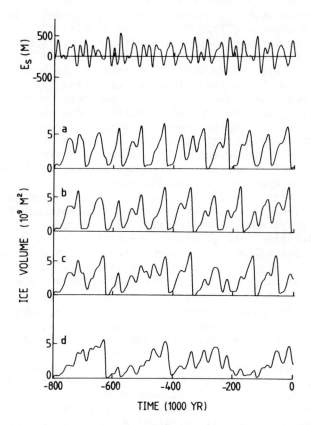

Figure 10.10. Results from integrations with a numerical
model of the northern hemisphere ice sheets. The upper
curve gives the imposed variations in snow-line
elevation. Maximum ice-accumulation rates are: 0.4
(run a), 0.375 (run b), 0.35 (run c), 0.25 (run d), in
m ice depth / yr. In run d the slope of the snow line
was slightly decreased. From Oerlemans (1982).

up and down in proportionality to the summer insolation variations (at 65 °N) caused by the orbital changes. This affects the model ice sheet in two ways. First, the mass balance varies in a way as described by (10.3.1). Second, the temperature at the ice surface, which serves as boundary condition in calculating the temperature distribution in the ice sheet, will change. The results shown all refer to experiments in which the snow-line temperature was set to -14 °C. For more details, the reader is referred to the original paper.

The upper curve in Figure 10.10 shows the forcing, i.e. the variations in snow-line elevation, for the last 800 000 yr (values of the insolation were taken from Berger, 1978). The other curves give the ice volume as a function of time, for various values of the maximum ice-accumulation rate appearing in (10.3.1).

First of all we note that in each run glacial cycles of long duration and asymmetric shape (large growth time, small decay time) occur. Clearly, the variance in the simulated ice-volume curves is on a larger time scale than that in the forcing. Although rapid deglaciations always occur during insolation maxima, the model ice sheet appears to have considerable internal freedom. Runs a, b and c are very similar in character, but differ with respect to their phase. Here the nonlinearities in the model, in particular the effect of basal melting on the ice-mass discharge, play an important role. Whether at a particular moment the melting point at the base is reached or not, may be decisive for the further evolution of the ice sheet.

The sensitivity of the simulated ice-volume curves to the ice-accumulation rate is a manifestation of a more general sensitivity to parameter variability. Small changes in the mean snow-line elevation, the geothermal heat flux, the snow-line temperature, etc., lead to similar differences in the phase of the simulated ice ages. The conclusion obviously should be that the dynamics of large ice sheets add a stochastic component to the climate system with a long time scale.

In view of this large sensitivity to various parameters it is virtually impossible to tune the model in such a way that the correlation between simulated and 'observed' ice-volume records is very high (apart from the problems associated with accurate dating of the deep-sea record). It is also questionable whether this is very useful, just like it does not make sense to tune a general circulation model of the atmosphere such that a one month's evolution of the world-wide flow pattern is reproduced. The important point is that ice-sheet dynamics provide a possible explanation for the pleistocene glacial cycles. In this interpretation, ice sheets were the dominating component of the climate system during this period.

10.5 Further remarks

As noted in the beginning of this chapter, many ice-age theories
have been proposed and it is only in the last few decades that
the importance of the physics of ice sheets has been recognized.
Even today theories are proposed in which ice sheets play a very
passive role and follow the climatic environment instantaneously.
On the other hand, more and more studies with ice-sheet models
are now carried out, and subtle instability mechanisms are being
incorporated. Such studies may lead to important new results, but
the basic mechanism will probably remain the same. Let us therefor
summarize the main findings and place them in broader perspective.
    The first thing to recall is that ice-sheet initiation in the
northern hemisphere is a very critical process. This follows
directly from the present state as well as from the fact that
during the last 100 million years or so the presence of large
ice sheets was exception rather than rule. In other words, the
northern hemisphere ice sheets have been close to the bifurcation
point (see Figure 8.5) for a long time. As a consequence, the
growth time of an ice sheet is not determined by the typical
ice-accumulation rate above the snow line as is sometimes argued,

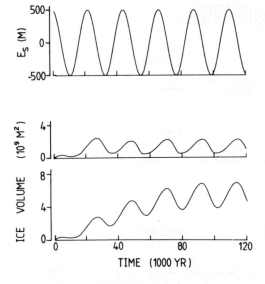

Figure 10.11. Response of a northern hemisphere ice-
sheet model to periodic forcing (20 000 yr). Snow-line
elevation is shown in the upper curve. Thermodynamics
were not included in the calculation. In the second run
(lower curve), the mean snow-line elevation was lowered
by 100 m with respect to the first run (middle curve).

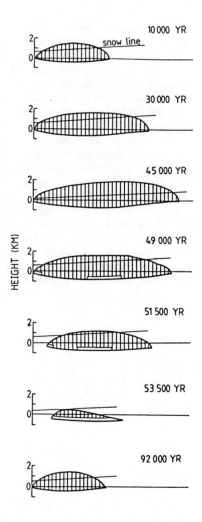

Figure 10.12. Ice-sheet profiles from an integration with periodic forcing. The presence of basal melt water is indicated. The period of the forcing is 20 000 yr, that of the response apparently about 80 000 yr for the particular set of model parameters used.
Simulated time is indicated in the pictures, note that they are not equidistant in time. In this particular run, a large value of the flow parameter was combined with a very small value of the slope of the snow line (to give an ice sheet that reaches ice-age size).

but by its mean value (over 10 000 yr, say) at a specific
(most favourable) location. A typical growth rate for ice
thickness can therefore easily be one tenth of the typical
accumulation rate. This is illustrated in Figure 10.11.

Weertman (1961) already stressed the fact that a large ice
sheet may destabilize itself, and recent calculations (as
discussed in this book, among other) support his arguments.
Bedrock sinking and heat accumulation are probably the most
important factors, although we cannot be sure that processes
involving the circulation in ocean and atmosphere are less
significant.

The crucial point really is that bedrock sinking and heat
accumulation lag the ice thickness. Consequently, typical decay
times can be very short (5000 to 10 000 yr). A model integration
in which this is very apparent is shown in Figure 10.12. Again,
the snow line moves up and down with a period of 20 000 yr.
After 49 000 yr of simulated time melt water is formed under the
central part of the ice sheet, and from that point the decay
proceeds very rapidly. From this example it is very clear that
lagged bedrock sinking brings a large part of the surface of
the ice sheet below the snow line.

No matter which process is most effective in destabilizing the
ice sheet, the result can be interpreted as an equivalent rise
of the snow line, and the typical behaviour of a northern
hemisphere ice sheet is best summarized by trajectories in a
steady-state solution diagram. In Figure 10.13 (see next page)
equilibrium states for a northern hemisphere ice sheet are
shown by heavy lines (compare the discussion in section 8.2).
When periodic forcing is imposed and destabilizing mechanisms are
absent, the ice volume will grow until a steady oscillation
settles down (an essentially linear response to the forcing).
This situation is depicted in the left-hand side of Figure 10.13.
The thin line shows the trajectory followed by the ice sheet.

When destabilizing effects are coming into play, and this is
'translated' into a progressive increase in snow-line elevation,
the picture changes completely. Due to the asymmetry in the
mass-balance field (ablation can be much larger than accumulation),
the ice-sheet is able to pass the bifurcation point during the
time that the snow-line elevation is large, as shown in the
right-hand side of the figure.

It should also be mentioned that the generation of long
glacial cycles is not very dependent on the particular form
of the forcing, as long as the bulk of its variance is on
shorter time scales than the 'natural' period for a glacial
cycle. Random movement of the snow line of sufficient strength
may also generate response on very long time scales ! In this
way the 'ice-dynamics theory' of the ice ages is a robust
theory: no matter where radiation variations come from, as long as
they are large enough glacial cycles will be produced.

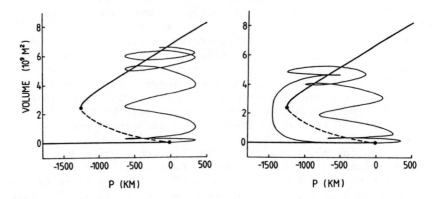

Figure 10.13. Trajectories (thin lines) of a northern
hemisphere ice sheet, subject to periodic forcing, in
the ice volume - climate point plane. Varying the
climate point is equivalent to moving the snow-line up
and down. Heavy lines represent equilibrium states.
The picture at the right refers to the case in which
the ice sheet is destabilized.

We conclude this chapter by a warning on the interpretation
of the oxygen-isotope record. The one-to-one relation between
the $0^{18}/0^{16}$ -record and global ice volume should not be taken
for granted. When conditions become colder the isotopic
composition of the ice sheets of Greenland and Antarctica
gradually changes, and the 'background-$0^{18}/0^{16}$ -ratio' will thus
not be constant. The magnitude of this effect can be up to $\frac{1}{4}$ of
the glacial-interglacial signal (Olausson, 1981).

It has also been suggested (Budd, 1981) that the 100 000 yr
power in the oxygen isotope record is generated by fluctuations
in Antarctic ice volume. Concerning the recent part of the
Pleistocene, the interpretation is that Antarctic ice volume
started to grow at 120 000 yr before present, at a slow rate,
and partly collapsed during the last deglaciation in the
northern hemisphere (15 000 yr ago). However, much more is needed
to be known about the dynamics of the Antarctic Ice Sheet before
such possibilities can be seriously investigated.

11. THE ICE SHEETS OF GREENLAND AND ANTARCTICA

In the preceding chapters we have extensively discussed the
theory of ice flow, thermodynamics of ice sheets, numerical
modelling, and in general terms the interaction of ice sheets
and climate. This should be the proper background to have a closer
look now at the ice sheets of Greenland and Antarctica, being
the only presently existing large ice sheets.

11.1 Short description

Compared to the total volume of the oceans, the ice volume stored
at Greenland and Antarctica is rather small, namely about 2.1 %.
Nevertheless, complete melting of all ice would change the
appearance of the earth' surface in a dramatic way. This becomes
evident if the ice volume is expressed in equivalent sea-level
rise (all other things being equal):

> East Antarctic Ice Sheet: 60 m
> West Antarctic Ice Sheet: 6 m
> Greenland Ice Sheet   :   6 m
> remaining ice   :   0.3 m

Note that melting of ice shelves, sea ice, and grounded ice
below sea level does not affect sea level.
   Apparently, the bulk of the earth's ice mass is stored at
Antarctica. Figure 11.1 (next page) gives an impression of the
difference in size between the Antarctic and Greenland Ice Sheet.
The set of maps published by the Scott Polar Research Institute
contains the most up-to-date information (Drewry, 1983).
A detailed description of the present-day ice sheets falls
outside the scope of this book, but some knowledge of the broad
physical characteristics is certainly required for the further
discussion on response to environmental changes.
   The Antarctic continent, about twice as large as Australia, is
almost completely covered by ice. Only about 1 % of the surface
is ice free, mainly in the Antarctic Peninsula and near the Ross
Sea, see Figure 11.2. The continent is naturally divided into
two parts, separated by the Transantarctic Mountains. The East
Antarctic Ice Sheet rests on a bedrock that would almost completely

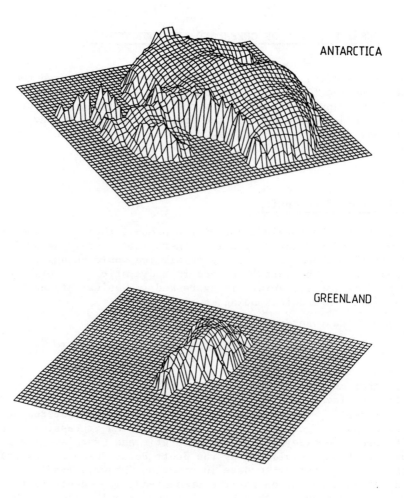

Figure 11.1. Three-dimensional plots of Greenland and Antarctica, on a 100x100 km grid.

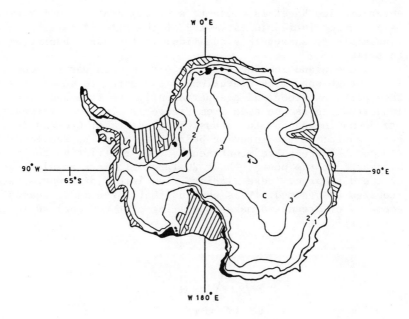

Figure 11.2. The Antarctic Ice Sheet. Isolines give
surface elevation (km above sea level). Black regions
are essentially free of ice. Ice shelves are indicated
by shading.

above sea level in case of no ice load. In contrast, a large
part of the West Antarctic Ice Sheet rests on a bed which is far
below sea level. Even in case of complete removal of the ice mass
and subsequent rising of the bedrock, a large part of the bed
would still be below sea level. This type of ice sheet has been
termed a marine ice sheet. Such a marine ice sheet is generally
surrounded by ice shelves, which can be very large when embayments
are present. Large ice shelves are found in the Ross and Weddel
Sea, see Figure 11.2.
    The surface of Antarctica is rather smooth, the maximum height
being about 4 km above sea level. The largest ice thickness is
found in a subglacial trench in Terre Adelie, namely almost
4800 m. Mean ice thickness on West and East Antarctica has the
same order of magnitude, but the surface of West Antarctica is
much lower and more irregular due to the rugged bedrock
topography.
    Accumulation rates are very low over the central part of East
Antarctica (about 3 cm ice depth/yr), but in coastal regions
accumulation reaches values up to 60 cm ice depth/yr. Melting is
insignificant, so ice mass is lost by calving only. Whether the

Antarctic Ice Sheet is close to a steady state is not known. The large uncertainties in estimates of the rate of calving make it impossible to answer this question. In the next chapter we return to this.

The Greenland Ice Sheet is subject to much warmer conditions. The average snow line elevation is estimated to be about 1500 m. Some parts of Greenland are permanently ice free and even inhabited. As can be seen from Figure 11.3, these bare parts are found in the coastal areas. The bedrock of Greenland is bowl-shaped: in the interior bedrock is close to sea level, while the coastal areas are mountaneous, thus restricting the ice-mass discharge to the ocean. About 20 large glaciers regulate the bulk of the discharge. These glaciers flow into fjords where large icebergs are formed. It is obvious that the complex coastal geometry of Greenland makes modelling of the Greenland Ice Sheet very difficult.

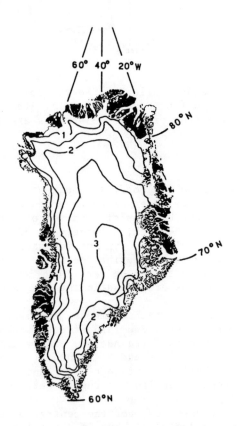

Figure 11.3. The Greenland Ice Sheet. Again, ice-free land is black. Isolines give surface elevation in km.

	GREENLAND	ANTARCTICA
area (10^6 km^2)	1.7	14
mean ice thickness (m)	1530	2160
ice volume (10^6 km^3)	2.6	30
maximum surface elevation (m)	3300	4000
annual accumulation (cm ice depth/yr)	34	17
loss of ice by calving	50 %	99 %
melting	50 %	1 %

Table 11.1. Some characteristics of the ice sheets of Greenland and Antarctica.

The mean annual accumulation is about twice as large as on Antarctica, namely about 34 cm ice depth. Ablation and production of icebergs are equally responsible for the loss of ice, but again it is not clear whether the total mass balance of the Greenland Ice Sheet is negative or positive.

The major features of the ice sheets of Greenland and Antarctica are summarized in Table 11.1. Apart from these major ice sheets, small ice caps are found on some Canadian and Soviet arctic islands. The largest of these are located on Ellesmere Island, where three ice caps cover an area of more than 20 000 km^2.

11.2   Ice shelves and ice streams

Drainage of the West Antarctic Ice Sheet largely takes place through ice streams which flow into ice shelves. In these streams discharge is essentially by sliding through troughs in the bed. The ice streams are separated by regions with slowly moving ice, probably frozen to the bedrock. Figure 11.4 illustrates such a configuration. It shows the situation in the Ross Sea region, where ice streams are very pronounced.

The fast moving ice streams (typical sliding velocity: 100 m/yr) are retarded by the drag exerted by the slowly moving ice on either side. Ice thickness in the ice streams is less than in the slowly moving ice. The ice streams determine to a large extent the profile of an ice sheet. Changes in environmental conditions will influence the flow properties of the ice streams, and the reaction of the ice sheet can be quite unexpected.

Another aspect concerns ice shelves that run aground in embayments and upset the normally smooth stress field. The presence of ice rises (occurring where an ice shelf runs aground) indicates that the ice-mass discharge from the grounded ice sheet to the ice front at the edge of the ice shelf is obstructed. The resulting backward pressure will of course lead to upstream thickening of the ice shelf and, as a consequence, to seaward migration of the grounding line.

The Ross Ice Shelf is presently the largest ice shelf with an area of about 525 000 km^2, which is one third of the total area covered by ice shelves around Antarctica. Its thickness varies from 1000 m at some places at the grounding line to about 200 m at the ice front. Ice velocities in the ice shelf are an order of magnitude larger than those in the ice sheet, ranging from a few hundreds of meters per year at the grounding line to several kilometers per year at the edge. Major supply is from the grounded ice sheet, although snow accumulation at the surface is not negligible. It is believed that melting at the bottom of the ice shelf can be as large as 1 m ice depth/yr.

Because ice shelves are floating and have direct contact with the ocean, they are susceptible to climatic change. Mercer (1978)

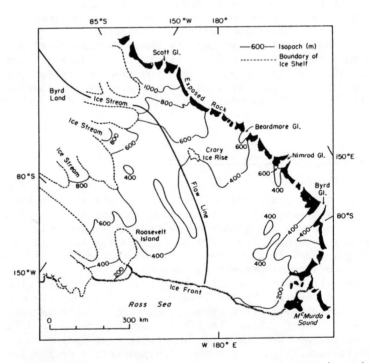

Figure 11.4. A map of the Ross Ice Shelf, showing lines of equal ice thickness (in m) and ice streams. From Paterson (1980), originally prepared by Robin (1975).

has stressed this point with regard to the CO_2-problem. He states
that when a climatic warming would occur, melting at the surface
and bottom of ice shelves will increase and may lead to the
disappearence of the major ice shelves of the West Antarctic Ice
Sheet. The lack of backward pressure then would cause the collapse
of the main part of this ice sheet. At present there is not much
evidence that we are at the threshold of such a dramatic event.
Some indication exist that the Ross Ice Shelf upstream of Crary
Island is actually thickening at a rate of about 0.3 m/yr, see
Thomas (1976).

Since the future of the West Antarctic Ice Sheet seems to be
closely tied to what happens with the ice shelves, it is desirable
to have a closer look at what happens at the grounding line. In
the next section we will deal with factors that may cause a
migration of the grounding line.

For more detailed descriptions of the ice sheets of Greenland and
Antarctica, and their ice shelf / ice stream systems, see Denton
and Hughes (1981).

11.3   Grounding-line migration

The grounding line is defined by the flotation criterion. If H is
the ice thickness at the grounding line (corrected for the less
dense firn layer at the surface), h_b the depth of the bed below
sea level, ρ and ρ_w the density of ice and water, respectively,
the following relation holds:

$$(11.3.1) \qquad h_b = H\ \rho/\rho_w.$$

Any change in ice thickness will of course lead to migration of
the grounding line. When for instance the ice sheet will become
thicker due to an increasing mass balance or decreasing ice
temperature, the grounding line will advance. A decreasing mass
balance, or thinning of the ice shelf, will result in retreat of
the grounding line.

It is important to realize, however, that generally a change
in the location of the grounding line, associated with a change
in environmental conditions, will be the result of a number of
competing effects. When leaving one out, almost any 'desired'
result can be obtained. As an illustration, consider the case
that climate becomes colder. Lower air temperatures will probably
lead to lower snow aacumulation, so the ice sheet will tend to
thin. After some time, however, ice temperature will start to
decrease, and the resulting lower creep rates will lead to
increasing ice thickness. For the ice shelf ocean temperature is

also important. Lower melting rates at the bottom of the ice shelf
thus compete with lower snow accumulation at the surface. Again,
strain rates will change due to the cooling. Adding to this the
fact that sea level may also change, it is obvious that the
ultimate migration of the grounding line is generally very
difficult to predict.

The distance over which the grounding line moves is strongly
influenced by the topography of the bed. This is illustrated in
Figure 11.5. If the bedrock in the vicinity of the ice sheet -
ice shelf junction slopes downward to the sea, migration of the
grounding line will be over comparatively short distances. The
steeper the slope of the bed, the smaller the grounding line
migration will be.

On West Antarctica the situation is more like the one depicted
in Figure 11.5b. The bed slopes upward to the sea, making the
response to changing environmental conditions (accumulation,
melting at the bottom of the ice shelf, but also changes in sea
level) much more complicated. A marine ice sheet, resting on a
bowl-shaped bed and extending to the edge of the continental
shelf seems to be a potentially unstable configuration.
From simple qualitative reasoning it seems that the ice sheet can
only attain two stable steady states, namely, one in which the
grounding line extends to the edge of the continental shelf, and
one in which the ice sheet has withdrawn to places where the bed
is above sea level.

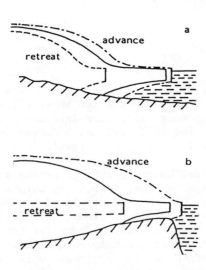

Figure 11.5. Effect of bedrock topography on the
migration of the grounding line. From Thomas (1979).

Simple and more sophisticated flow-line models of the type
discussed in section 4.4 have been used to study how vulnerable
the West Antarctic Ice Sheet is to changes in sea level and
the rate of thinning in the Ross Ice Shelf (e.g. Young, 1981;
Thomas et al., 1979; Lingle and Clark , 1984). Although such
flow-line studies certainly give insight in the type of processes
that are important, it is doubtful whether a reliable answer
concerning the West Antarctic Ice Sheet can be obtained. The
problem is really two-dimensional, as can be seen in Figure 11.4.
The Ross Ice Shelf is fed by the ice streams coming down from the
West Antarctic Ice Sheet, but also by the glaciers originating
in the Transantarctic Mountains. Each catchment area may react
in a different way to climatic change, and it is likely that the
two-dimensional flow in the Ross Ice Shelf has to be considered
explicitly in order to arrive at conclusions about grounding-line
migration in the Ross Sea.

In the ice streams sliding is very important. Near the
grounding line sliding velocities are an order of magnitude
larger than deformational velocities. It is thus not surprising
that grounding-line retreat or advance in flow-line models
depends very strongly on the particular formulation used to relate
basal sliding to the stress field.

It is well-known that the sliding velocity depends on the
driving shear stress as well as the normal load (Lliboutry, 1968;
Weertman, 1972; Reynaud, 1973). A recent experimental study on
sliding was carried out by Budd et al. (1979). Their experiments
showed that in general a sliding law of the type

$$(11.3.2) \qquad V_s = c \, \tau_b^n \, / \, N^m$$

holds, where V_s is the sliding velocity, τ_b the basal shear stress,
and N the normal load. When basal water at pressure P is present,
the normal load should be corrected for the upward buoancy force,
i.e. N has to be replaced by N-P.

Budd et al. found that for low stresses and normal loads (lower
than actually occurring in ice sheets) $n \simeq 1$ and $m \simeq 1$. For larger
τ_b and N, it appeared that $n \simeq 3$ and $m \simeq 1$. However, fitting a sliding
law of the type (11.3.2) to observed slding velocities on West
Antarctica suggests $n \simeq 1$ and $m \simeq 2$, with $c = 5 \times 10^6$ bar^{-1} m^3 yr^{-1}
(Budd, personal communication).

In view of the previous discussion it is clear that the
reduction of normal load by buoyancy effects is mainly responsible
for the high sliding velocities observed in the West Antarctic
(and other) ice streams. When basal water is present, and
'connected to the sea', the reduced normal load is

$$(11.3.3) \qquad N^* = \rho g \, (b - \frac{\rho}{\rho_w} H) \quad .$$

H denotes ice thickness, and b elevation of the bed with respect
to sea level. So according to (11.3.2) and (11.3.3), sliding
velocities will become very large when the grounding line is
approached, and as a consequence the slope of the ice surface
(and thus τ_b) will be small.

Opinions about the present state of the West Antarctic Ice
Sheet differ widely. It is generally accepted that during the
last ice age the Ross Sea was completely covered by grounded ice,
and that the grounding line retreated to its present position in
holocene time. However, whether grounding-line retreat is still
continuing is not known. Hughes (1973) suggested that the West
Antarctic Ice Sheet is still disintegrating. Estimated values of
grounding-line retreat in the Ross Sea are typically 50 to 100
m/yr since the last glacial maximum. Thomas and Bentley (1978),
on the other hand, think that retreat stopped about 7000 yr ago.
With a quantitative model of grounding-line migration they arrived
at the result that the grounding line in the south-eastern part
of the Ross Sea is actually advancing. This could be the result
of lagged bedrock rebound following the removal of the ice-age
ice cover (see also Greischar and Bentley, 1980). Although
estimates of the rate of grounding-line migration are quite
large (up to 100 m/yr), verification from direct measurements is
very difficult. It is obvious that more sophisticated models and
more detailed measurements are needed to attack the problem of
the stability of the West Antarctic Ice Sheet.

11.4 A model of the Antarctic Ice Sheet

With regard to modelling of the Antarctic Ice Sheet and its
response to climatic variations, the previous section is not very
encouraging. Neverheless, when interest is in climatic states that
differ widely (by a 10 K difference in mean temperature, say),
the subtle dynamics of the grounding line in the Ross Sea, and
probably also in the Weddell Sea, may be less important. In this
section we discuss a model of the Antarctic Ice Sheet with
relatively simple dynamics, but designed in such a way that it
can be run for any prescribed mean sea-level temperature thought
to represent climatic conditions in the antarctic. The model
described below was presented in Oerlemans (1982a).

The evolution of the Antarctic Ice Sheet is calculated on a
100x100 km grid, laid out over the bedrock topography shown in
Figure 11.7. This topography was obtained from the present-day
bed as given in Atlas Antarktiki (1966), corrected for the
present loading by assuming isostatic balance. So the figure
shows the estimated equilibrium bedrock topography in case of no
ice load.

Ice flow is computed with the model described in section 4.1,
with the numerical scheme (4.1.4)-(4.1.5) generalized for the

two-dimensional case. So in this model the flow does not depend
on ice temperature or normal load, but is a function of the
driving shear stress τ_b. Nevertheless, since ice shelves probably
play an important role when conditions on West Antarctica are
marginal for ice-sheet growth, it is useful to include a
pragmatic scheme for the simulation of ice shelves. We employ
a formulation of the type

$$(11.4.1) \qquad H_{sh} = \frac{1}{Q} \int_0^{2\pi} H(\lambda)/Y^p(\lambda) \ d\lambda \quad .$$

This equation states that at a particular point the thickness
H_{sh} of the shelf is determined by the ice thickness at the
grounding line (H), averaged over direction (λ) and weighted with
the distance to the grounding line to some power p. Q is a
normalization constant, depending on the particular discrete form
with which (11.4.1) is replaced for use on a grid. In the present
model only 4 directions were 'checked'. The exponent p can be
adjusted to give realistic thicknesses, a reasonable value is
p=1.5. The major reason that (11.4.1) works is that large ice
shelves are only formed in embayments, and that apparently the
degree of enclosure by grounded ice is more important than the
details of the ice-shelf thinning process.

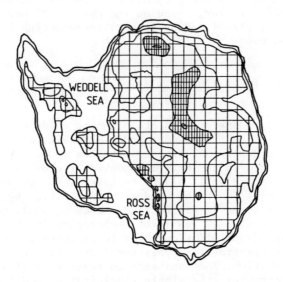

Figure 11.7. Topography of the Antarctic continent in
case of no ice load. The lowest contour is -2 km. Light
shading: above sea level; heavy shading: over 2 km.

When in the model ice shelves have been calculated, a simple
check is made to see whether floating ice runs aground at some
place. If so, at this place ice flow is then treated according to
the flow law employed for grounded ice. This procedure allows
growth of an ice sheet in the centre of a shallow sea due to
ice shelf-grounding.

The reaction of the bedrock to the ice load is calculated from
simple damped return to isostatic equilibrium, with a relaxation
time of 5000 yr. Although this is only a first approximation to
the actual reaction of the earth's mantle, it seems to be
sufficiently accurate for a global model. Note, by the way, that,
for a given distribution of ice, the equilibrium state of the
bed does not depend on the flow in the asthenosphere. The
difference is in the time needed to reach this state.

Since we want to apply the model to a wide range of climatic
conditions, care has to be taken in formulating the mass balance.
As discussed earlier, snow accumulation depends on such factors
as surface slope, elevation, air temperature and distance to the
moisture source (see the discussion of section 9.2). For a given
sea-level temperature T_{sl}, the annual accumulation is parameterized
as (in m ice depth/yr)

$$(11.4.2) \qquad P = \max\{0.05; \ (0.3 + 14\ S - 4\times10^{-5}\ h)/C\} \ .$$

The surface elevation is denoted by h, its slope by S. The constant
C is a continentality factor, based on the number of surrounding
grid points (in a circle of radius R, say) representing ice cover
or bare ground. We use

$$(11.4.3) \qquad C = \min\{1 + N/N_{max}; \ 2\} \ .$$

So C runs from 1 for a one-grid point island to 2 for the
continental interior. In the present model R is set to 500 km.

In (11.4.2), the dependence of precipitation on air temperature
is implicitly included in the h-term. However, the overall effect
of sea-level temperature still has to be taken into account. This
is done by multiplying the right-hand side of (11.4.2) by a
function $f(T_{sl})$, as shown in Figure 11.8. Snow accumulation is
largest when the annual mean sea-level temperature is around the
freezing point, and drops off slowly for lower temperatures and
sharply for higher temperatures. This concludes the formulation
of the snow accumulation. The constants introduced in (11.4.2)
were obtained by matching this simple precipitation model with
the observed accumulation pattern over Antarctica (according to
Atlas Antarktiki, 1966). The function $f(T_{sl})$ was derived from
a mixture of climatic data on snowfall in Canada and Scandinavia,

as well as output from sensitivity experiments carried out with
general circulation models of the atmosphere.

Although at present melting on Antarctica is negligible, it
may be important in warmer climates.To derive a relation between
melt rates and annual surface temperature, measurements from the
Greenland Ice Sheet (Ambach, 1972) can be used. The result is
(again in m ice depth/yr, see Figure 11.8a):

$$(11.4.4) \qquad \text{Melt} = \max \{0; \ 0.028(12+T_s)^2\} \ , \ T_s > -12 \ ^{\circ}C \ .$$

T_s denotes annual mean surface temperature, which can be calculate
from T_{sl} by using a lapse rate. Eq. (11.4.4) is not valid for
temperatures above 0 $^{\circ}$C, but this is not serious because in that

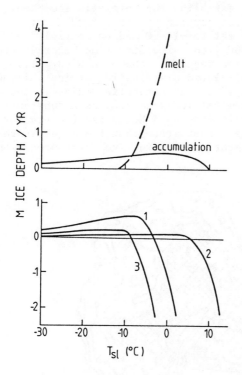

Figure 11.8. Mass-balance parameterization for the
Antarctic Ice Sheet model. In the upper panel ablation
and accumulation are shown as a function of annual
sea-level temperature, for zero elevation and surface
slope. The lower panel shows the mass balance for
(1): S=0.02, h=500 m, C=1.5; (2): S=0, h=2000 m, C=1;
(3) S=0, h=0, C=1. From Oerlemans (1982a).

case ice cover does not occur anyhow.

Combining (11.4.2)-(11.4.4) now gives a complete description of the mass balance in terms of sea-level temperature, surface slope, surface elevation, and continentality. A few typical resulting relations between mass balance and sea-level temperature are shown in Figure 11.8b. Obviously, differences are very large. For the continental interior at high elevation (curve 3), the mass balance is small but remains positive for high annual sea-level temperatures. The largest values of the mass balance are found on the ice-sheet edges, for annual sea-level temperatures around -5 °C.

This parameterization of the mass balance completes the model of the Antarctic Ice Sheet. The imposed climatic conditions are formulated in terms of one single parameter, namely, annual sea-level temperature around the Antarctic continent.

A first test of this model is to see whether it is able to produce an ice sheet resembling the Antarctic Ice Sheet as we observe it today. An integration was therefore carried out with sea-level temperature set to -14 °C and no initial ice mass. It then takes about 30 000 yr of simulated time before a steady state is approached. The result is shown in Figure 11.9.

In general, the calculated distribution of ice thickness is very reasonable, except in the Antarctic Peninsula and Weddell Sea regions. Here the model ice sheet covers a substantially larger area. Since it is not known how close the present ice sheet is to equilibrium, one cannot state that the model is really in error. A similar integration starting from present-day bedrock

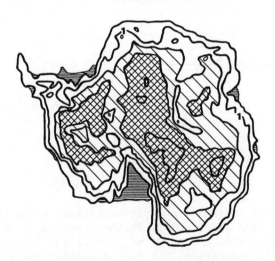

Figure 11.9. Steady-state ice thickness calculated with the model for present conditions. Thickness over 3 km is shown by cross-hatching. From Oerlemans (1982a).

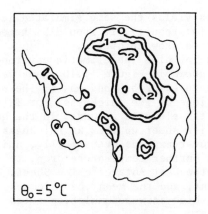

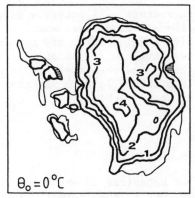

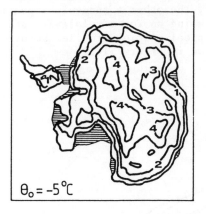

Figure 11.10. The Antarctic Ice Sheet in warmer climates. Annual sea-level temperature is indicated in $^{\circ}C$. The contour interval for ice thickness is 1000 m, and ice shelves are indicated by shading. Oerlemans (1982a).

topography and ice thickness yields the same simulated ice sheet
(this is by no means a trivial result). It should be noted that
the Ross and Amery ice shelves show up nicely.

Results from experiments in which the climatic environment of
Antarctica was changed are shown in Figure 11.10. Runs were made
for annual sea-level temperatures of 5,0 and -5 ^{o}C. The pictures
show steady-state distributions of ice thickness. For T_{sl}=5 ^{o}C,
ice cover is restricted to the elevated grounds. In the higher
part of East Antarctica an ice sheet up to 2 km thickness has
formed. For T_{sl}=0 ^{o}C, ice cover expands dramatically, and a few
minor ice shelves appear. A further temperature drop, to -5 ^{o}C,
leads to the initiation of the West Antarctic Ice Sheet. Now
large ice shelves are present, and the mean ice thickness is quite
large, a consequence of the high snow accumulation rates (see
Figure 11.8a).

More subtle sensitivity experiments will probably give unre-
liable results, because both the effect of ice temperature on ice
flow and the impact of extensive basal sliding are not included.
The influence of changes in sea level on the Antarctic Ice Sheet,
for instance, can only be investigated when these factors are
dealt with in the model. It is probably fair to say that this
level of modelling has not yet been reached.

11.5 Surging of ice sheets

Some glaciers exhibit a type of behaviour in which two alterna-
ting flow regimes are apparent. One is the 'normal' steady movement of ice down the valley, the other a suddenly appearing rapid
ice-mass discharge (surge) going on for a relatively short time
(months to a few years). Typical glacier profiles before and just
after such an event are shown in Figure 11.11.

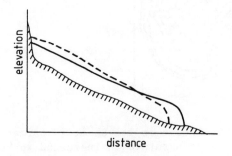

Figure 11.11. Typical glacier profiles before (dashed
line) and after a surge (solid line).

An adequate explanation for glacier surging has not yet been given. An overview of current theories can be found in Paterson (1981). Most of them relate the transient behaviour to varying thermal conditions at the bed (in particular, water lubrication).

It has been suggested (Wilson, 1964, 1969; Hollin, 1972, 1980) that large ice sheets also surge. This idea is based on proxy data of sea-level stands, indicating that in the past sudden (100 to 1000 yr) rises in sea level occurred. Hollin (1980) discusses the evidence pointing to a major Antarctic surge 95000 yr ago.

Although a number of physical models have been developed to study the dynamics of surging glaciers (e.g. Campbell and Rasmussen, 1969; Clarke, 1976; Budd, 1975), only very few studies consider the problem of ice-sheet surging. Budd and McInnes (1979) applied their model for self-surging glaciers to East Antarctica.

The model of Budd and McInnes is based on the assumption that changes in frictional heating is the major source for variations in basal temperature. As a direct consequence, basal melting only occurs when the ice-mass discharge exceeds some critical value. If the mass flux is larger than this value basal sliding sets in, and the additional frictional heating keeps the process going on for a while. The surge is actually transmitted through the glacier by longitudinal stress gradients. When the ice thickness in the upper part of the glacier has been decreased, the ice-mass discharge and thus the dissipative heating slows down, and the glacier comes into a stagnant mode. Gradual built up then paves the way for a new surge. This self-sustained oscillation owes its

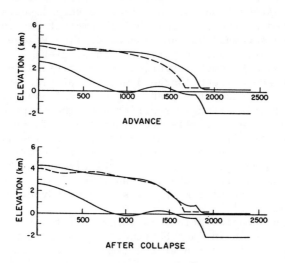

Figure 11.12. Ice-sheet profiles before and after a surge, as calculated by Budd and McInnes (1979) for an

existence to the dependence of basal water production (although
not calculated explicitly in the Budd and McInnes model) on the
ice-mass discharge, and vice versa.

Figure 11.12 shows a result obtained by Budd and McInnes for
an East Antarctica flow line. For the particular model constants
they used, the period of oscillation is about 23 000 yr, with
minor surges in between. The major surge, associated with the
profiles shown in the figure, takes about 250 yr. The fact that
surging along a flow line on East Antarctica may occur is a very
interesting result, but the question has to be raised whether
this particular model is really applicable to large ice sheets.

The main problem involves the thermodynamics. In the Budd and
McInnes model the coupling between temperature field and ice flow
is in fact formulated by defining a reduced basal shear stress τ^{*}:

$$(11.5.1) \qquad \tau^{*} = \tau/(1+\tau\phi V) \quad .$$

Here V is the mean ice velocity and ϕ the so-called friction
lubrication factor (the essential tuning parameter). Basal
sliding is then directly related to the reduced shear stress.

A detailed description of the Budd and McInnes model is
outside the scope of this book, but on the basis of (11.5.1) a
few remarks can be made. The term $\tau\phi V$ fromulates the model's
'thermodynamics'. Although this is a reasonable approach for
a valley glacier, where frictional heating dominates the heat
budget, it seems to be too simple for ice sheets. As discussed
extensively in chapters 5 and 6, advection is very important. This
is best illustrated by the fact that basal temperatures in an ice
sheet go down when the mass balance increases (see Figures 6.4
and 6.7). In contrast, basal temperatures in glaciers increase
when the mass balance becomes larger, as parameterized in (11.5.1).

Another difference between 'surge dynamics' of glaciers and
ice sheets is that in case of a glacier a well-defined flow line
exists. The ice flow is strongly restricted by the shape of the
valley, and therefore essentially one-dimensional. In case of an
ice sheet it is difficult to see that a particular flow line
will remain so. Streamlines are generally not trajectories.
Another effect involves advection in lateral direction. This
will try to damp an eventual surge, because ice from higher
elevations with lower basal temperatures will be advected towards
the 'central surge line'.

In view of these considerations it seems fair to state that
ice-sheet surging should in fact be studied with two-dimensional
models, with a full calculation of the temperature field. This
is very (computer) time consuming, and also the procedure
developed by Budd and McInnes to deal with the normal stresses
is not so easily generalized for the two-dimensional case.

However, even if normal stress are neglected self-sustained

oscillations of an ice sheet are possible when the interaction
of ice-mass discharge and ice temperature is taken into account.
To illustrate this possibility, we now discuss some results from
a pilot study we carried out with the Antarctic Ice Sheet model
discussed in section 11.4, in which the temperature calculation
of section 5.4 was implemented. The effect of temperature on ice
flow is as formulated in sections 6.3 and 6.4, i.e. the flow
parameter depends on the basal temperature and amount of basal
water. Again basal water spreads in a diffusive way, now with
diffusivity linearly proportional to normal load. The flow
parameter B is a piecewise linear function of basal water W:

(11.5.2)

$$B = WfB_0 \quad \text{if} \quad 0 \leqslant W \leqslant 1 \text{ m} ,$$

$$B = fB_0 \quad \text{if} \quad W \geqslant 1 \text{ m} .$$

B_0 is the flow constant for a basal temperature at the melting
point, and f a basal sliding constant determining the maximum
increase in ice-mass discharge due to extensive sliding.

As expected, the behaviour of the model strongly depends on the
value of f. For small values of f, the model ice sheet grows to
a steady state with very large ice volume. Large values of f
lead to oscillatory behaviour. Ice-thickness over East Antarctica
then varies periodically, while the West Antarctic Ice Sheet
settles to a steady state with extensive basal melting and high
ice velocities toward the Ross Sea.

Figure 11.13 shows the result for f=30 m^{-1}. In this case the

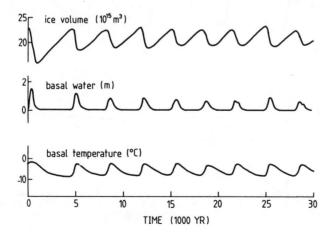

Figure 11.13. A periodic solution of the Antarctic Ice
Sheet model, for f=30 and present-day climatic conditions.

period of oscillation is short (about 3000 yr). Note that the
model ice sheet needs time to 'get in balance'. The mean ice
volume is considerably smaller than the one given in Table 11.1
for the Antarctic Ice Sheet. The reason is that in the model
experiments 'old' data on bedrock elevation were used, while
the estimate given in the Table is based on recent measurements
of the bedrock topography (Drewry, 1982). These modern data
reveal a lower mean bedrock elevation, and consequently a larger
estimate of ice volume (30×10^{15} m^3).

   A noteworthy observation is that a surge is not initiated by a
gradual warming of the ice sheet (this is evident from the
temperature curve), but by accumulation of basal water in the
thicker parts, ultimately reaching the coastal regions. By that
time the runaway ice-mass discharge begins. So it is quite
obvious that the occurrence of surges in the models depends
critically on how the spread of basal water is parameterized.

   Figure 11.14 gives ice volume curves for two other values of f.
As the one shown in Figure 11.13, these curves apply to runs
with the present state as initial condition. For f=20 m^{-1}
oscillations of longer period and larger amplitude occur, while
for f=8 m^{-1} the model ice sheet grows to a steady state. In all
cases, the West Antarctic Ice Sheet hardly responds to what
happens on the eastern part of the continent, and stays in the
basal melting – sliding mode.

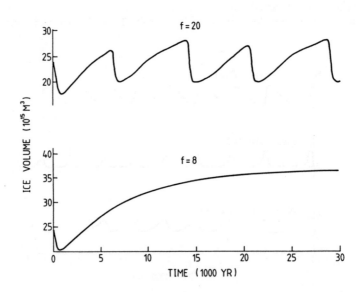

Figure 11.14. Ice volume versus time for model runs
with f=20 and f=8 m^{-1}.

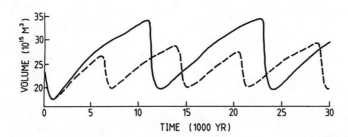

Figure 11.15. The effect of sea-level temperature on
periodic solutions. The dashed line is the f=20 run
from Figure 11.14, the solid line shows the result for
a 5 K temperature drop.

Figure 11.15 finally shows the sensitivity of the model result to
sea-level temperature. The dashed curve is the one for f=20 in
Figure 11.14, and the solid curve shows an integration in which
annual sea-level temperature was dropped by 5 K. This turns out
to have a dramatic effect. The surge amplitude now amounts to 40%
of the total ice volume, the period of the oscillation being
10 000 yr.

Altogether, both approaches for studying surging of ice sheets
presented here suggest that such events are very well possible.
The most critical factor seems to be whether basal water can
accumulate. If not, that is, if all water prodeced is removed
instantaneously for all conditions, surging is unlikely to occur.
Experiments with the Antarctic Ice Sheet model described here,
in which f=0 but the effect of ice temperature on deformational
velocity was included, revealed that then the feedback between
dissipation and ice flow is far too weak to generate cyclic
behaviour. So creep instability is not a likely mechanism to
explain ice-sheet surging.

12. ICE SHEETS IN THE FUTURE

One of the reasons that the dynamics of ice sheets are studied is
the desire to obtain knowledge about their future behaviour. In
particular during the last few million years ice volume on earth
varied enormously. Cold periods with huge ice sheets on the
northern hemisphere continents were interrupted by relatively warm
interglacials, in which only the ice sheets of Greenland and
Antarctica survived. From a purely statistical point of view, it
is very likely that a new ice age is on its way.
   The influence of human activities on climate is another point
of interest (and concern). Of all possible effects the increase
in atmospheric CO_2-concentration due to the use of coal, oil and
gas is presumably most important. The associated climatic warming
will certainly have its influence on the evolution of the major
ice sheets.
   In this chapter we consider a few of these issues, after a
brief discussion on how natural variability and transient behaviour
may obscure the response of the cryosphere to a change in
climatic conditions.

12.1   The signal-to-other effects ratio

Ice sheets have a long time scale, making it very difficult to
measure directly any changes in their shape. In most cases, the
growth or shrinkage of an ice sheet is the result of an imbalance
between accumulation and ablation (and/or calving) of may be one
percent. So to see whether an ice sheet is actually growing or
shrinking, very accurate measurements are required. It is unlikely
that such measurements become available in the near future.
   Even if we could determine that an ice sheet grows, it is not
so obvious that we can trace the cause. The climate system is a
very complex, highly nonlinear system in which various types of
time-dependent behaviour, (quasi-)periodic or chaotic, may occur
without any change in external forcing. Ice-sheets will of course
react to "internal" climatic variability, as discussed in a highly
simplified way in section 8.3.
   A useful concept in assessing whether a specific evolution is
meaningful in the sense that it can be attributed to a specific

195

change in steady forcing rather than to natural variability is
the signal-to-noise ratio. Various definitions of this quantity
exist, but the basic idea is the same. Figure 12.1 shows two time
series of some climatic element X(t). Both series have the same
trend, but in series A the trend is certainly not statistically
significant. The signal-to-noise ratio should be defined in such
a way that it clearly discriminates between curves A and B.

A current procedure is to subtract the trend and the mean from
the original series, i.e. to write

$$(12.1.1) \qquad X'(t) = X(t) - \bar{X} - \alpha(t-T/2) \quad .$$

Here $\bar{X}$ is the mean value of X, α is the trend, and t runs from
t=0 to t=T. Note that $\bar{X}'=0$. The signal-to-noise ratio S can now
be defined as

$$(12.1.2) \qquad S = \int_0^T \alpha^2 (t-T/2)^2 dt / \int_0^T X'^2(t)dt = \frac{1}{12} \alpha^2 T^3 / \int_0^T X'^2(t)dt$$

So S is nothing but the ratio of the variance contained in the
trend to the variance in all other (that is, smaller) time scales.
The value of S required to call a specific trend significant is
not only a matter of statistics, but also of taste.

Questions concerning the impact of man-made climatic perturbations
on the present-day ice sheets, in particular the CO_2 production,
have led to a number of speculations. In many studies, the present
state of the ice sheets is intuitively assumed to be close to

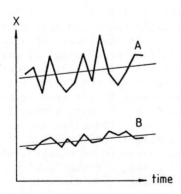

Figure 12.1. Two time series with the same trend, but
with different noise levels.

equilibrium, and any significant climatic change due to human
activities would pull the ice sheets out of equilibrium. However,
this is not a very realistic way of looking at the problem. When
accumulation rates are small, and they are over large parts of
the Antarctic Ice Sheet, the response time is very large. The
effect of ice temperature and bedrock sinking makes the response
even slower. For an ice accumulation rate of 5 to 10 cm ice/yr,
the response time can easily be 30 000 yr or more. In view of
this, it is very unlikely that the Antarctic Ice Sheet, which
started to change its shape dramatically when the last ice age
came to en end, is close to an equilibrium state.

So although the expected CO_2 warming, when it finally shows
up, will undoubtly affect the state of the Greenland and
Antarctic ice sheets, it is not clear whether this signal will
be larger than changes in ice volume that would occur anyhow.
This remark is not to discourage modellers studying the possible
effect of a CO_2 warming on the cryosphere, but to draw attention
to the very long memory of ice sheets.

. When a model of the Antarctic Ice Sheet, with considerable
internal freedom, is integrated in time with the present ice
distribution as initial condition, it will certainly drift away
from the present state (if not, the conclusion should probably
be that the model is too heavily constrained rather than that
it simulates the ice sheet well). A general procedure in such
a situation is to compare a climatic perturbation experiment with
a control run. The difference between perturbation and control
experiment that gives some indication of how the system might
react. When it turns out that this difference is much smaller
than the difference between control experiment and present
observed state, the effect of the climatic perturbations is
unimportant, or the model does a poor job.

12.2   The carbon dioxide problem

Carbon dioxide is a natural constituent of the atmosphere,
although the concentration is small. Estimates of the atmospheric
CO_2 content at the beginning of the industreal era vary between
270 and 300 ppmv (part per million by volume). Since 1958
accurate measurements have been carried out at various widely
separated locations. These measurements show that the CO_2 content
of the atmosphere is now increasing rapidly, see Figure 12.2.

Future CO_2 concentrations depend on the rate of emission,
absorption by biomass and take-up by the oceans. Although many
aspects of the global CO_2 cycle are poorly understood, it is
not considered likely that the CO_2 content of the atmosphere
will more than double in the next 100 years. Probably the
increase will be somewhat smaller.

CO_2 is very important with regard to the radiation balance of

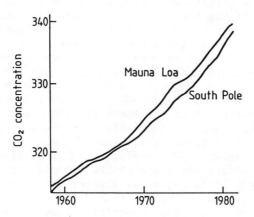

Figure 12.2. Carbon dioxide concentration as measured
at the Mauna Loa (Hawaii) ans South Pole observatories.
The measurements have been smoothed in time. From
Smith (1982).

the atmosphere. Qualitatively, the effect of more CO_2 in the
atmosphere can be understood with the aid of the simple radiation
model described in the first section of this book. Increasing
CO_2 content leads to a larger emissivity of the atmosphere, and
due to the associated increase in counter radiation surface
temperature will go up.

    Calculations with more refined radiation models, e.g.
Ramanathan (1981), indicate that on doubling the atmospheric
CO_2 content a global warming of 2 to 5 K should be expected,
depending on the specific assumptions concerning cloud, water
vapour and albedo feedback. General circulation models have
also been used to study the climatic implications of increasing
carbon dioxide concentration (e.g. Manabe and Stouffer, 1982).
The results reveal a considerable seasonal and spatial variability
in the response. A general result is that the warming at high
latitudes is a factor two to three larger than the global mean
effect. The question of how the cryosphere reacts to an increasing
CO_2 concentration in the atmosphere is thus very relevant.

    The effect of a warmer atmosphere on ice sheets acts through
two different processes. In the first place the mass balance
will change, bringing the ice sheet immediately out of balance.
Secondly, higher air temperature implies that the ice sheet will
gradually warm up. However, for the flow parameter to increase,
higher temperatures must have reached the deeper layers of the
ice sheet. This requires thousands and thousands of years. We
now first turn to the mass-balance effect.

The mass balance effect will not be the same for the Greenland
and Antarctic ice sheets. The reason for this becomes clear when
we consider Figure 12.3. Here the ice sheets of Greenland and
Antarctica are drawn in a mass-balance field, assumed to
represent general conditions. The most important feature is that
at some elevation the mass balance reaches a maximum value.
Above this level, atmospheric moisture content decreases rapidly
(see Figure 9.3) and accumulation rates become small. In a first
approximation, a climatic warming implies a uniform upward shift
of the mass-balance field. It is then easy to see that the
consequences for the Greenland and Antarctic ice sheets are
different. The total mass balance of the Antarctic Ice Sheet
will increase, whereas that of the Greenland Ice Sheet will
decrease.

A quantitative result for the Antarctic Ice Sheet can be
obtained with the model discussed in section 11.4. Assuming an
increase in surface temperature of 3 K and in precipitation of
12 % during the next 100 yr (which mimics the effect of a
doubling CO_2 content according to Manabe and Stouffer, 1980),
leads to an increase of the Antarctic ice volume of about 0.5 %
after 250 yr. All other things being equal, this corresponds to
a sea-level drop of approximately 0.3 m. Calculations for the
Greenland Ice Sheet (e.g. Ambach, 1980) indicate a substantial
decrease in its mass balance. For a similar scenario, ice volume
on Greenland would decrease by 3 % in the next 250 yr. The
corresponding rise in sea-level would be about 0.2 m. These
estimates thus indicate that there will be a net increase in
global ice volume due to the mass-balance effect. It is important
to realize that this only applies to the immediate future. The
decreasing mass balance of Greenland could ultimately lead to a

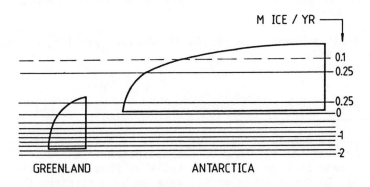

Figure 12.3. Illustrating the different environmental
conditions of the Greenland and Antarctic ice sheets.
Horizontal lines show the mass-balance field.

complete collapse of the ice sheet, while the Antarctic Ice
Sheet reaches a new steady state with slightly larger ice volume.

As stated earlier, it takes a long time before the basal layers
of an ice sheet feel a temperature change at the surface. Young
(1981) studied this process for a flow line in West Antarctica
debouching into the Ross Ice Shelf. He found that, starting from
an increase in surface temperature of 5 K, it takes 500 to 1000
yr before a new thermal equilibrium is reached in the thinner
parts of the ice shelf. For the interior of the ice sheet this
time is of the order of 20 000 yr. When the ice shelf reaches
a new equilibrium, the thinning rates are increased by about
5 cm/yr. Young thus concluded that within the next few centuries
the temperature effect on the ice flow is very small.

Finally attention should be paid to the effect of increasing
ocean temperature. The warming of the ocean mixed-layer will
be of the same order of magnitude as the increase in surface
air temperature, and this will certainly effect the ice shelves
around Antarctica. Mercer (1978) has stressed this point. He
(and others) have suggested that a slight increase in the
thinning rate of the Ross Ice Shelf (and also of the Ronne Ice
Shelf in the Weddell Sea) will lead to a situation in which the
ice shelves do not run aground anymore at a number of places
where they do now. The backward pressure on the main ice sheet
would then be reduced substantially, eventually leading to a
complete collapse of the West Antarctic Ice Sheet. At present,
model calculations exist that support this view, whereas other
do not. The situation is so complicated that present models are
not yet capable of handling the dynamics in a reliable way. All
statements concerning an eventual collapse of the West Antarctic
Ice Sheet due to a CO_2 warming should therefore be considered
as speculative. Better models are needed to change this
situation.

For further reading on the carbon dioxide problem, we recommend
Smith (1982).

12.3 The next ice age

Turning to a much longer time scale, we now consider the question
of the next ice age. A mere extrapolation of the climatic record
as we know it from proxy data tells us that another ice age is
coming. In the chapter on ice ages we have stressed the
possibility of a large stochastic component. Simulated ice-volume
curves run out of phase easily, due to the nonlinear effects in
the response of ice sheets to environmental conditions.

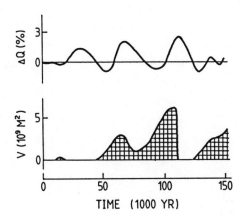

Figure 12.4. Future caloric summer insolation variations for 65 °N (according to Berger, 1978a), and associated simulated ice volume (V).

In spite of this, it is obvious that the probability of ice sheet inception and growth is larger when insolation at high latitudes is low. Let us therefore have a look at the Milankovitch insolation variations as calculated for the future by Berger (1978). Again, we consider the deviation of caloric summer halfyear insolation at 65 °N, see Figure 12.4. The next minimum, at 20 000 yr after present, is weak and of short duration, and probably not significant enough to initiate a major glaciation. Another, stronger minimum is found at 50 000 yr after present, and is more likely to lead to ice-sheet growth in the northern hemisphere. This is suggested by some experiments we carried out with the ice-sheet model employed in section 10.4 to simulate the proxy ice-volume record.

Experiments with sets of model parameters fitting well the ice-volume record of the last few hundred thousands of years (in particular the last two major terminations) all predict an ice volume curve for the future as shown in Figure 12.4. This result should be considered as a global indication, however.

To find out which factors are decisive to the inception of large ice sheets in the northern hemisphere, much more detailed investigations are needed. In particular the dependence of the mass balance on such factors as atmospheric circulation, orography, ocean temperature, etc., needs further attention, and here many possibilities for modelling exist.

EPILOGUE

Ice-sheet modelling is a challenging field, and we are only at the beginning of its development. Studying the role ice sheets play in the climate system requires interdisciplinary work, and we hope that this book has provided some background and ideas. Emphasis has been on modelling aspects, not on field work and its direct interpretation. Extensive descriptions of former and present day ice sheets, and the traces they left, can be found in Denton and Hughes (1981). This book also contains an exhaustive list of references.

Another point we did not discuss is the excellent capability of ice sheets to register some climatic parameters. Oxygen isotope ratios of the ice give information on temperature conditions that once prevailed (e.g. Dansgaard et al., 1977), dust concentrations and acidity may tell something about extension of deserts or volcanic activity (e.g. Hammer et al., 1980), total gas content of the ice gives indications on former surface elevation (e.g. Raynaud and Whillans, 1982), etc. It is quite obvious that such measurements from ice cores can play a very important role in the validation of ice sheet models.

We like to make two points on model verification. Firstly, verification should preferably be done by comparing a time-dependent simulation to a proxy data record. Running a model to equilibrium and comparing the result to a presently observed state is a much less powerful procedure. Secondly, it is a matter of good science philosophy to explain a particular mechanism or series of event in the simplest possible way. Once a specific result has been obtained, it is extremely useful to find out how far the model can be simplified while still producing the same result.

Finally, this book was on processes with long time scales, and one should not speak too dramatically about the impact of cryospheric changes on human activities.
The world's major problem is the arms race, not the next ice age or the possible collapse of the West Antarctic Ice Sheet.

REFERENCES

Abramowitz M and Stegun I A (1965), Handbook of mathematical
functions. Dover Publ. Inc., New York.

Ambach W (1972), Zur Schätzung der Eis-Nettoablation im
Randgebiet des Grönlandischen Inlandeises. Polarforschung 42,
18-23.

Ambach W (1980), Anstieg der CO_2 Konzentration in der Atmosphäre
und Klimaänderung: Mögliche Auswirkungen auf den Grönlandischer
Eisschild. Wetter und Leben 32, 135-142.

Atlas Antarktiki (1966). Glawnoje Oeprawlenie Geodezii i
Kartografii MG, Moskwa-Leningrad.

Berger A L (1975), The astronomical theory of paleoclimate: a
cascade of accuracy. In: Proceedings of the WMO-IAMAP
symposium on long-term climatic fluctuations. WMO Publ. No.
421, 65-72.

Berger A L (1978), Long-term variations of daily insolation and
quaternary climatic changes. J. Atmos. Sci. 35, 2362-2367.

Berger A L (1978a), Long-term variations of caloric insolation
resulting from the earth's orbital elements. Quaternary Res.
9, 139-167.

Birchfield G E, Weertman J and Lunde A T (1981), A paleoclimate
model of northern hemisphere ice sheets. Quaternary Res. 15,
126-142.

Bodvarsson G (1955), On the flow of ice sheets and glaciers.
Jokull 5, 1-8.

Broecker W S, Thurber D L, Goddard J, Ku T, Matthews R K, and
Mesolella K J (1968), Milankovitch hypothesis supported by
precise dating of coral reefs and deep-sea sediments.
Science 159, 1-4.

Brouwer D, and Van Woerkom A J J (1950), Secular variations of
the orbital elements of principal planets. Astron. Papers
Am. Ephemeris 13, 81-107.

Bryan K, Manabe S, and Pacanowski R C (1975), A global ocean-
atmosphere climate model. Part II: the oceanic circulation.
J. Phys. Ocean. 5, 30-46.

Budd W F (1970), The longitudinal stress and strain-rate gradients in ice masses. J. Glaciology 9, 19-27.

Budd W F (1975), A first simple model for periodically self-surging glaciers. J. Glaciology 14, 3-21.

Budd W F (1981), The importance of ice sheets in long term changes of climate and sea level. IAHS Publ. No. 131, 141-471.

Budd W F, Jenssen D, and Radok U (1971), Derived physical characteristics of the Antarctic Ice Sheet. ANARE Glaciology Publ. No. 120, Melbourne.

Budd W F, Keage P L, and Blundy N A (1979), Empirical studies of ice sliding. J. Glaciology 23, 157-170.

Budd W F and McInnes B J (1979), Periodic surging of the Antarctic ice sheet - an assessment by modelling. Hydr. Sci. Bull. 24, 95-104.

Budd W F, Young N W, and Austin C R (1976), Measured and computed temperature distributions in the Law Dome Ice Cap, Antarctica. J. Glaciology 16, 99-109.

Campbell W J and Rasmusson L A (1969), Three-dimensional surges and recoveries in a numerical glacier model. Canadian J. Earth Sci. 6, 979-986.

Clark J A, Farrell W E, and Peltier W R (1978), Global changes in postglacial sea level: a numerical calculation. Quaternary Res. 8, 265-287.

Clarke G K C (1976), Thermal regulation of glacier surges. J. Glaciology 16, 231-250.

Clarke G K C, Nitsan U, and Paterson W S B (1977), Strain heating and creep instability in glaciers and ice sheets. Rev. Geoph. Space Phys. 15, 235-247.

CLIMAP (1976), The surface of the ice-age earth. Science 191, 1131-1137.

Dansgaard W, Barkov N I, and Splettstoesser J (1977), Stable isotope variations in snow and ice at Vostok, Antarctica. IAHS Publ. No. 118, 204-209.

Dansgaard W, Johnsen S J, and Clausen H B (1977), Stable isotope profile through the Ross Ice Shelf and Little America V, Antarctica. IAHS Publ. No. 118, 322-325.

Denton G H and Hughes T J , eds., (1981), The last great ice sheets. John Wiley, New York.

Drewry D J (1982), Antarctica unveiled. New Scientist, 244-251.

Drewry D J, ed., (1983), Antarctica: glaciological and geophysical folio. Scott Polar Research Institute, Cambridge.

Flint R F (1971), Glacial and quaternary geology, John Wiley,
    New York.

Ghil M and Le Treut (1981), A climate model with cryodynamics
    and geodynamics. J. Geophys. Res. 86, 5262-5270.

Gilmore R (1981), Catastrophe theory for scientists and engineers.
    John Wiley, New York.

Greischar L L and Bentley C R (1980), Isostatic equilibrium
    grounding line between the West Antarctic inland ice sheet
    and the Ross Ice Shelf. Nature 283, 651-654.

Hammer C U, Clausen H B, and Dansgaard W (1980), Greenland ice
    sheet evidence of post-glacial volcanism and its climatic
    impact. Nature 288, 230-235.

Hays J D, Imbrie J, and Shackleton N J (1976), Variations in the
    earth's orbit: pacemaker of the ice ages. Science 194,
    1121-1132.

Heath C R (1979), Simulations of a glacial paleoclimate by three
    different atmospheric general circulation models. Paleogr.,
    -clim., -ecology 26, 291-303.

Hollin J T (1972), Interglacial climates and Antarctic ice surges.
    Quaternary Res. 2, 401-408.

Hollin J T (1980), Climate and sea level in isotope stage 5: an
    East Antarctic ice surge at 95 000 BP ? Nature 283, 629-633.

Holton J R (1972), An introduction to dynamic meteorology.
    Academic Press, New York.

Hoskins B J and Karoly D J (1981), The steady linear response of
    a spherical atmosphere to thermal and orographic forcing.
    J. Atmos. Sci. 38, 1179-1196.

Houghton J T (1977), The physics of atmospheres. Cambridge
    University Press, Cambridge.

Hughes T (1973), Is the West Antarctic Ice Sheet disintegrating ?
    J. Geophys. Res. 78, 7884-7910.

Imbrie J and Imbrie J Z (1980), Modeling the climatic response
    to orbital variations. Science 207, 943-953.

Imbrie J and Imbrie K P (1979), Ice ages: solving the mystery.
    MacMillan Press, London.

Imbrie J, Van Donk J, and Kipp N G (1973), Paleoclimatic
    investigation of a late pleistocene Carribean deep-sea core:
    a comparison of isotopic and faunal methods.

Jaeger J C (1969), Elasticity, fracture and flow. Third ed.,
    Methuen, London.

Jenssen D (1977), A three-dimensional polar ice-sheet model.
    J. Glaciology 18, 373-390.

Källen E, Crafoord C, and Ghil M (1979), Free oscillations in a climate model with ice-sheet dynamics. J. Atmos. Sci. 36, 2292-2303.

Kominz M A and Pisias N G (1979), Pleistocene climate: deterministic or stochastic ? Science 204, 171-173.

Kukla G, Berger A L, Latti R, and Brown J (1981), Orbital signatures of interglacials. Nature 290, 295-300.

Lamb H H (1977), Climate. Present, past and future. Vol. 2. Climatic history and the future. Methuen, London.

Lau N C (1979), The observed structure of troposheric stationary waves and the local balances of vorticity and heat. J. Atmos. Sci. 36, 996-1016.

Lingle C S (1984), A numerical model of interactions between a polar ice stream and the ocean: application to ice stream E, Antarctica. J. Geophys. Res., in press.

Lliboutry L A (1968), General theory of sub-glacial cavitation and sliding of temperate glaciers. J. Glaciology 7, 21-58.

Lliboutry L A (1979), Local friction laws for glaciers: a critical review and new openings. J. Glaciology 23, 67-95.

Lockwood J G (1979), Causes of climate. Edward Arnold, London.

Lorenz E N (1967), The nature and theory of the general circulation of the atmosphere. WMO Publ. No. 218, Geneva.

Manabe S and Stouffer R J (1980), Sensitivity of a global climate model to an increase of CO_2 concentration in the atmosphere. J. Geophys. Res. 85, 5529-5554.

Mercer J H (1978), West Antarctic ice sheet and CO_2 greenhouse effect: a threat of disaster. Nature 271, 321-325.

Mesinger F and Arakawa A (1976), Numerical methods used in atmospheric models. Vol. 1. WMO, GARP Publ. No. 17, Geneva.

Neumann G and Pierson W J (1966), Principles of physical oceanography. Prentice-Hall, Englewood Cliffs, New Jersey.

North G R and Coakley J A (1979), Differences between seasonal and mean annual energy balance calculations of climate and climate sensitivity. J. Atmos. Sci. 36, 1189-1204.

Nye J F (1959), The motion of ice sheets and glaciers. J. Glaciology 3, 493-507.

Oerlemans J (1980), On zonal asymmetry and climate sensitivity. Tellus 32, 489-499.

Oerlemans J (1980a), Continental ice sheets and the planetary radiation budget. Quaternary Res. 14, 349-359.

Oerlemans J (1980b), Some model studies on the ice-age problem. Royal Netherlands Meteorological Institute, Scientific Report No. 158, De Bilt.

Oerlemans J (1980c), Model experiments on the 100 000 yr glacial cycle. Nature 287, 430–432.

Oerlemans J (1982), Glacial cycles and ice-sheet modelling. Climatic Change 4, 353–374.

Oerlemans J (1982a), A model of the Antarctic Ice Sheet. Nature 297, 550–553.

Oerlemans J (1983), A numerical study on cyclic behaviour of polar ice sheets. Tellus 35, 81–87.

Oerlemans J and Van den Dool H M (1978), Energy-balance climate models: stability experiments with a refined albedo and updated coefficients for infrared emission. J. Atmos. Sci. 35, 371–381.

Oerlemans J and Vernekar A D (1981), A model study of the relation between northern hemisphere glaciation and precipitation rates. Contr. Atmos. Phys. 54, 352–361.

Olausson E (1981), On the isotopic composition of late cenozoic sea water. Geogr. Ann. 63A, 311–312.

Oort A H and Rasmusson E M (1971), Atmospheric circulation statistics, NOAA Prof. Pap. No. 5, U.S. Dept. of Commerce, Rockville.

Oort A H and VonderHaar T H (1976), On the observed annual cycle in the ocean-atmosphere heat balance over the northern hemisphere. J. Phys. Ocean. 6, 781–800.

Opsteegh J D and Van den Dool H M (1980), Seasonal differences in the stationary response of a linearized primitive equation model: prospect for longe-range weather forecasting ? J. Atmos. Sci. 37, 2169–2185.

Palmen E and Newton C W (1969), Atmospheric circulation systems. Academic Press, New York.

Paltridge G W and Platt C M R (1976), Radiative processes in meteorology and climatology. Elsevier, Amsterdam.

Paterson W S B (1980), Ice sheets and ice shelves. In: Dynamics of snow and ice masses (ed. Colbeck S C), Academic Press, New York, 1–78.

Paterson W S B (1981), The physics of glaciers. Pergamon Press, Oxford.

Pedlosky J (1979), Geophysical fluid dynamics. Springer Verlag, New York.

Peltier W R (1980), Models of glacial isostasy and relative sea level. In: Dynamics of plate interiors, American Geophys. Union, Geodynamics Ser. I, 111-128.

Peltier W R (1982), Dynamics of the ice age earth. Adv. Geophys. 24, 1-146.

Pollard D (1978), An investigation of the astronomical theory of the ice ages using a simple climate - ice sheet model. Nature 272, 233-235.

Pollard D (1980), A simple parameterization for ice sheet ablation rate. Tellus 32, 384-388.

Pollard D, Ingersoll A P, and Lockwood J G (1980), Response of a zonal climate - ice sheet model to the orbital perturbations during the quaternary ice ages. Tellus 32, 301-319.

Pollard D (1982), A simple ice sheet model yields realistic 100 kyr glacial cycles. Nature 296, 334-338.

Ramanathan V (1981), The role of ocean - atmosphere interactions in the CO_2 climate problem. J. Atmos. Sci. 38, 918-930.

Raynaud D and Whillans I M (1982), Air content of the Byrd core and past changes in the West Antarctic Ice Sheet. Ann. Glaciology 3, 269-273.

Reynaud L (1973), Flow of a valley glacier with a solid friction law. J. Glaciology 12, 251-258.

Robin, G de Q (1955), Ice movement and temperature distribution in glaciers and ice sheets. J. Glaciology 2, 523-532.

Robin, G de Q (1975), Ice shelves and ice flow. Nature 253, 168-172.

Saltzman B and Vernekar A D (1971), An equilibrium solution for the axially symmetric component of the earth's macroclimate. J. Geophys. Res. 77, 3936-3945.

Sanberg J A M and Oerlemans J (1983), Modelling of pleistocene European ice sheets: the effect of upslope precipitation. Geologie en Mijnbouw, in press.

Sanderson T J O (1979), Equilibrium profiles of ice shelves. J. Glaciology 22, 435-460.

Sanderson T J O and Doake C S M (1979), Is vertical shear in an ice shelf negligible ? J. Glaciology 22, 285-292.

Sellers W D (1965), Physical climatology. Univ. of Chicago Press, Chicago.

Shackleton N G and Opdyke N D (1973), Oxygen isotope and strati-graphy of equatorial Pacific core V28-238: oxygen isotope temperatures and ice volumes on a 10^5 and 10^6 yr time scale. Quaternary Res. 3, 39-55.

Shackleton N G and Opdyke N D (1976), Oxygen isotope and
paleomagnetic stratigraphy of Pacific core V28-239, late
Pliocene to latest Pleistocene. Geol. Soc. Am. Mem. 145,
449-464.

Smith G D (1978), Numerical solution of partial differential
equations: finite difference methods. Clarendon Press, Oxford.

Smith I N (1982), Carbon dioxide - emissions and effects.
Report ICTIS/TR 18, IEA Coal Research, London.

Stommel H (1948), The westward intensification of wind-driven
ocean currents. Trans. Amer. Geophys. Union 99, 202-206.

Thomas R H (1973), The creep of ice shelves: theory.
J. Glaciology 12, 45-53.

Thomas R H (1976), Thickening of the Ross Ice Shelf and
equilibrium state of the West Antarctic ice sheet. Nature 259,
180-183.

Thomas R H (1979), The dynamics of marine ice sheets.
J. Glaciology 24, 167-177.

Thomas R H and Bentley C R (1978), A model for holocene retreat
of the West Antarctic ice sheet. Quaternary Res. 10, 150-170.

Thomas R H, Sanderson T J O, and Rose K E (1979), Effect of
climatic warming on the West Antarctic ice sheet. Nature 277,
355-358.

Turcotte D L (1979), Flexure. Advances in Geoph. 21, 51-86.

Turcotte D L and Schubert G (1982), Geodynamics. Applications of
continuum physics to geological problems. John Wiley, New York.

Van den Dool H M (1980), On the role of cloud amount in an energy
balance model of the earth's climate. J. Atmos. Sci. 37,
939-946.

Van der Veen C J and Oerlemans J (1984), Global thermodynamics of
a polar ice sheet. Tellus, in press.

Vernekar A D (1972), Long-period global variations of incoming
solar radiation. Meteor. Mon. Vol. 12, Am. Meteor. Soc.,
Boston.

Walcott R I (1970), Isostatic response to lading of the crust in
Canada. Can. J. Earth Sci. 7, 716-727.

Walcott R I (1973), Structure of the earth from glacio-isostatic
rebound. Ann. Rev. Eartg & Plan. Sci. 1, 15-37.

Weertman J (1961), Equilibrium profiles of ice caps.
J. Glaciology 3, 953-964.

Weertman J (1961a), Stability of ice-age ice sheets. J. Geoph.
Res. 66, 3783-3792.

Weertman J (1968), Comparison between measured and theoretical temperature profiles of the Camp Century, Greenland, borehole. J. Geophys. Res. 73, 2691-2700.

Weertman J (1969), Water lubrication mechanism of glacier surges. Can. J. Earth Sci. 6, 929-942.

Weertman J (1972), General theory of water flow at the base of a glacier or ice sheet. Rev. Geophys. Space Phys. 10, 287-233.

Weertman J (1973), Creep of ice. In: Physics and chemistry of ice (Eds. Whalley E, Jones S J, and Gold L W), Royal Soc. of Canada, Ottawa.

Weertman J and Birchfield G E, Subglacial water flow under ice streams and West Antarctic ice sheet stability. Ann. Glaciology 3, 316-320.

Wilson A T (1964), Origin of ice ages: an ice shelf theory for pleistocene glaciation. Nature 201, 147-149.

Wilson A T (1969), The climatic effects of large-scale surges of ice sheets. Can. J. Earth Sci. 6, 911-918.

Young N W (1981), Responses of ice sheets to environmental changes. IAHS Publ. No. 131, 331-360.

INDEX

Ablation   134, 137-140
Accumulation   140-144, 175-177, 184-185
Activation energy   95
Albedo (feedback)   2-4, 12, 23-28, 140, 144, 147-152, 162
Antarctic Ice Sheet   15, 21, 102, 107, 126, 127, 137, 141, 154,
                        157, 173-193, 199-200
Asthenosphere   111-123
Astronomical theory   153-162
Atmospheric
   absorptivity   1-6
   emissivity   1-6
   transmissivity   4
   circulation   9-13, 171

Backward pressure   72-73, 179, 200
Baroclinic instability/waves   8-11, 36, 139
Barotropic vorticity equation   37
Basal
   melting   77, 87, 92-93, 101, 106-109, 191
   shear stress   49, 55-56, 60, 72-74, 181
   temperature   80-93, 97, 103-106, 165
Bedrock
   adjustment   103, 111-123, 163-165
   temperature   93
Bifurcation   25, 97, 100, 101, 106, 134, 169, 171
Boundary conditions for ice-sheet models   69-74
Brunhes-Matuyama reversal   154

Carbon dioxide problem   195, 197-200
Circulation theorem   29
Climate
   sensitivity   4-5, 27-28, 147-152
   system   1-19
Cloudiness   4, 23, 28, 139
Constitutive equations 46-47
Convection
   in atmosphere   3, 138
   in ocean   16
Coriolis force   7, 16, 29, 37
Creep function   42

213